GEOMETRICAL AND TECHNICAL DRAWING
FOR SECONDARY SCHOOLS

PARTS TWO AND THREE

GEOMETRICAL AND TECHNICAL DRAWING

a complete course in six parts

Book 1: Part One

Book 2: Parts Two and Three

Book 3: Parts Four, Five and Six

with *Teacher's Guide*

GEOMETRICAL AND TECHNICAL DRAWING

FOR SECONDARY SCHOOLS

SI METRIC EDITION

Parts Two and Three

BY

H. A. FREEBURY

With an Introduction by

W. GILLESPIE

M.Eng., B.Sc.(Eng.), M.I.Mech.E., M.I.Prod.E.,
A.F.R.Ae.S., A.M.I.Mar.E.

Principal: The College, Swindon

CASSELL · LONDON

CASSELL & COMPANY LTD
35 Red Lion Square · London WC1
and at
MELBOURNE · SYDNEY · TORONTO
JOHANNESBURG · AUCKLAND

© Cassell & Co. Ltd 1959, 1970
All Rights Reserved. No part of this publication may be reproduced, stored in a retrieval system, or transmitted, in any form or by any means, electronic, mechanical, photocopying, recording or otherwise, without the prior permission of Cassell and Company Ltd.

First published 1959
Second edition November 1959
Third edition August 1960
Fourth edition November 1960
Fifth edition March 1963
Sixth edition September 1964
Seventh edition February 1968
Eighth (revised) edition March 1970
Eighth edition, second impression April 1971
I.S.B.N. 0 304 93577 8

Throughout this book the following British Standard specifications have been followed:
BS 1991 (Part 1): 1961: *Letter Symbols, Signs and Abbreviations*
BS 308: 1964: *Engineering Drawing Practice*
BS 3763: 1964: *The International System (SI) Units*

Printed in Great Britain by
Lowe & Brydone (Printers) Ltd., London, N.W.10

Preface

This revised course in three books covers the G.C.E. examination in the subject at Ordinary Level as set by most of the Examining Bodies. Books 1 and 2 will also cover the Technical Drawing courses of the C.S.E. Regional Boards and other comparable organizations.

It is written with the continued hope of encouraging all groups in secondary schools. Being set down in some detail, it can be used almost as a self-educator by the brighter pupils, thus giving the instructor more opportunity to move around the class and help individually those in difficulty.

The diagrams have been kept as simple as the subject will allow and the author feels, in the light of considerable experience in secondary modern schools and an Evening Institute, that many exercises included are within the capacity of the lower groups. Such constructions as the American method of drawing isometric circles have been retained for this reason.

Since the course begins with fundamentals, it could be started in the second year, with the harder problems in each chapter of Part I left until later. It is suggested that the whole course might be taken over a period of three years, for frequent revision is essential to ensure that the more difficult constructions have been thoroughly grasped.

No attempt has been made to give formal proofs of the geometrical problems included, but the principles underlying the more difficult constructions have been stated where considered to be of help.

I must again express my gratitude to Mr. W. Gillespie, M.Eng., B.Sc.(Eng.), M.I.Mech.E., M.I.Prod.E., A.F.R.Ae.S., A.M.I.Mar.E., Principal of the College, Swindon, for his continued interest in this course and for the valuable suggestions he has made.

Finally, I must thank the Senate of London University and the Associated Examining Board for their permission to use questions from past examination papers in the Tests to be found after each Part, and to convert the dimensions on these to approximate metric equivalents.

H. A. FREEBURY

Contents

PART II. SOLID GEOMETRY

Chapter		Page
1.	The Solids	3
2.	Orthographic Projection	
	A. The Principle	6
	B. The Method	8
	C. Representation of Surfaces	9
	D. Representation of Prisms	14
	E. Representation of Pyramids	17
3.	Isometric Drawing	20
4.	Oblique Drawing	29
5.	Sections	
	A. Sections of Solids	33
	B. Sections of Hollow Bodies	42
6.	Auxiliary Views	
	A. Auxiliary Elevations	48
	B. Auxiliary Plans	52
	C. Auxiliary Sectional Views	54
	D. Auxiliary Views of a Straight Line	57
	(i) The Orthographic Method	58
	(ii) The Conical Method	62
7.	Intersection of Solids	67
8.	Development of Surfaces	73
	Test Papers	86

CONTENTS (continued)

PART III. TECHNICAL DRAWING

Chapter		Page
1.	Methods of Fastening	91
2.	Freehand Sketching	98
3.	Making the Drawing	
	A. The Layout	103
	B. Reading the Drawing	104
	C. The Method	105
4.	Sectional Elevations and Plans	112
5.	Third Angle Projection	119
	Test Papers	124

Note: All dimensions throughout this book are in millimetres unless otherwise stated.

Introduction

The teaching of Geometrical and Technical Drawing is one of growing importance in the Secondary and Grammar Schools, and it is most desirable that correct principles are observed from the outset of a course of drawing. In the past much work has been of reduced value to a student subsequently attending a Technical College because fundamental principles have not been established, and this work has had to be repeated before the work of the College can proceed.

Mr. Freebury has had considerable industrial experience which must add to the value of the book he has produced. Industrial experience cannot be over-estimated in the teaching and preparation of drawing. The methods adopted are sound and if his suggestions are followed a student coming to a Technical College will find the work progressive. He has produced an extensive cover of all aspects a student will need, which if properly assimilated will give an excellent knowledge of the subject.

W. GILLESPIE

Conversion Table

The following table may be of use where it is preferred to work any exercises in Imperial units. The equivalents are approximate only, but should be found satisfactory for this purpose.

Milli-metres	Inch equivalents		Approximate equivalents in fractions of an inch		
	Decimal	Fractions	Sixteenths	Tenths	Twelfths
1	0·04		0·6	0·4	0·5
2	0·08		1·3	0·8	0·9
3	0·12	$\frac{1}{8}$	1·9	1·2	1·4
4	0·16		2·5	1·6	1·9
5	0·20		3·2	2·0	2·4
6	0·24	$\frac{1}{4}$	3·8	2·4	2·8
7	0·28		4·4	2·8	3·3
8	0·32		5·0	3·2	3·8
9	0·35	$\frac{3}{8}$	5·7	3·5	4·2
10	0·39		6·3	3·9	4·7
12	0·47	$\frac{1}{2}$	7·6	4·7	5·7
14	0·55		8·8	5·5	6·6
15	0·59		9·4	5·9	7·1
16	0·63	$\frac{5}{8}$	10·1	6·3	7·6
18	0·71		11·3	7·1	8·5
20	0·79	$\frac{3}{4}$*	12·6	7·9	9·4
24	0·94		15·1	9·4	11·3
25	0·98	1	15·8	9·8	11·8
30	1·18	$1\frac{1}{4}$†	18·9	11·8	14·2
40	1·57	$1\frac{1}{2}$	25·2	15·7	18·9
50	1·97	2	31·5	19·7	23·6

* 19 mm is a more precise equivalent for $\frac{3}{4}$ in
† 32 mm is a more precise equivalent for $1\frac{1}{4}$ in

PART II
Solid Geometry

A Solid has Three Dimensions, Length, Breadth and Height.

CHAPTER ONE

The Solids

The solids which will be dealt with in this book are: Prisms, Pyramids, Cones and Spheres.

Fig. 101 shows two views—a front view and a plan view (looking on the top from above)—of the commoner geometrical solids.

(*a*) to (*g*) are prisms. A *prism* is a solid whose ends are equal and parallel plane figures, and whose sides form rectangles (or parallelograms). A prism is named according to the shape of its ends. The prisms we shall deal with are:

        Triangular (*a*)
        Square   (*b*)
        Rectangular (*c*)
        Pentagonal (*d*)

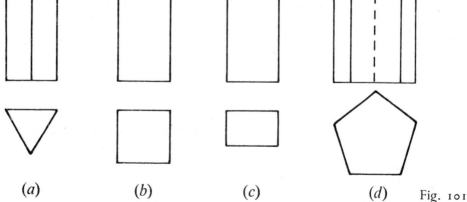

Fig. 101

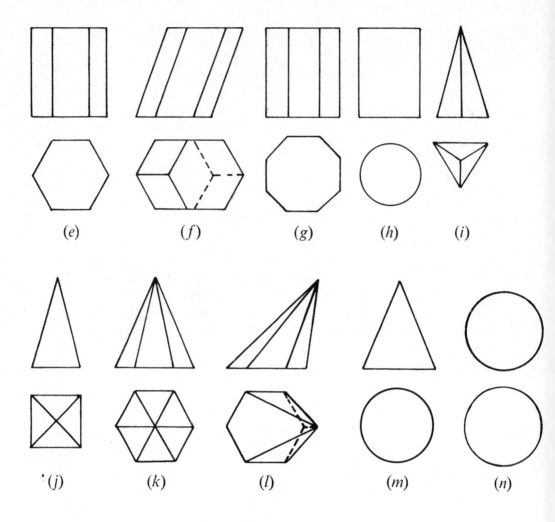

Fig. 101

Hexagonal   (e) and (f)
Octagonal   (g)

A cube is a prism with all its edges the same length, while the cylinder (h) is a *circular prism*.

A prism is called *regular* or *irregular* according to the shape of its ends, and if the axis—the imaginary line running through the centre of the solid—is at right angles to the base, the prism is known as a *right prism*. If the axis is not at right angles to the base, it is called an *oblique prism*. All the prisms

4

in Fig. 101 are right prisms with the exception of (*f*), which is an oblique reular hexagonal prism.

A *pyramid* is a solid consisting of a base and sloping triangular sides meeting at a point called the apex. A pyramid is also named according to the shape of its base, which can be regular or irregular.

The pyramids shown in Fig. 101 are:

        Triangular pyramid (*i*)
        Square pyramid   (*j*)
        Hexagonal pyramid (*k*) and (*l*)

If the axis is at right angles to the base, the pyramid is said to be a *right* one. If it is not at right angles to the base, the pyramid is called *oblique*, as at (*l*).

If some part of the pyramid is cut away, the pyramid is said to be *truncated*, and the remaining portion next to the base is called the *frustum*.

Cone (*m*) is the special name given to a circular pyramid; its surface is not composed of triangles. It can also be a *right* or *oblique* cone.

A *sphere* (*n*) consists of a solid formed by turning a semi-circle about its diameter. When it is cut into two equal parts, each half is known as a *hemisphere*.

CHAPTER TWO

Orthographic Projection

A. THE PRINCIPLE

In geometrical and technical drawing there are two general methods of showing solids. They are:

(a) Orthographic projection
(b) Pictorial drawing (e.g. isometric and oblique drawing).

Orthographic projection is most commonly used, and is a means whereby a solid can be shown on a plane surface such as the drawing paper.

In this type of drawing the object under consideration is imagined to be suspended in space, in front of and above three planes arranged at right angles to each other as shown in Fig. 102. These planes are known as the *vertical plane*, the *horizontal plane*, and the *side vertical plane*.

From the object, sight lines are cast in the direction of arrow 'A' on to the vertical plane, in the direction of arrow 'B' on to the horizontal plane, and in the direction of arrow 'C' on to the side vertical plane. This gives a different outline on each of the planes, as you can see by studying Fig. 103.

Now imagine these planes to be hinged screens, each containing an impression of the object as shown, so that the side vertical plane with its impression can be rotated through 90° along OZ until it forms one plane with the vertical plane, as shown at Fig. 103.

Now let these two planes, vertical and side vertical, be rotated as one through 90° till they form one single plane with the horizontal plane, as in Fig. 104. We now have three impressions of the object in one plane. These impressions (or views) are called: *front elevation*—or often just *elevation*—being the one on the vertical plane; *plan*, being the one on the horizontal plane; and *end elevation*, being the one on the side vertical plane.

The word 'elevation' here means 'view', and must not be confused with altitude or height.

The above is the principle of the orthographic projection known as *first*

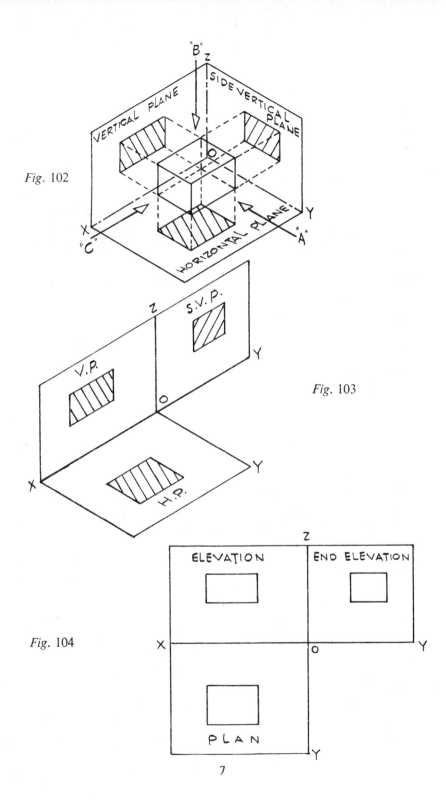

Fig. 102

Fig. 103

Fig. 104

angle. There is also an orthographic projection known as *third angle*, and this will be discussed in Part III, Chapter 5.

Remember the following points:
1. The bend line XOY (Fig. 104) is usually known as the *ground line*.
2. The views should be in projection. That is, they should be the same distance above and below XOY and each side of ZOY, with the plan vertically in line with the front elevation, and the end elevation in line with the front elevation.
3. As this is first-angle projection, the views should always be arranged as shown, with the plan beneath the elevation and the end elevation to the right of the front elevation.

Finally, it is usual to refer to the three planes of reference as V.P. (vertical plane), H.P. (horizontal plane) and S.V.P. (side vertical plane). This reference will be adopted for brevity in the rest of this book.

B. THE METHOD

Fig. 105(*b*) shows a pictorial drawing of the block shown in 105(*a*), but slightly enlarged.

Fig. 105(*a*) shows the three orthographic views of the same block. For the moment, ignore the lettering round these views, while the method of constructing them is detailed.

To set out the complete drawing, draw the elevation first. Next, the hinge lines XOY and ZOY are drawn a convenient distance from the elevation. Lines called *projection lines* or *projectors*, and shown here as thin lines, are drawn from the vertical edges of the elevation to obtain the length of the plan. The upper edge of the plan view is then drawn the same distance below XOY (in future referred to as either simply XY or the ground line) as the base line of the elevation is above it. The length of the block is marked on this line from the vertical projectors and the view is then completed.

To obtain the end elevation, projectors are drawn from the horizontal edges of the plan to the hinge line ZOY or ZY. The points P and Q are transferred to the ground line either by means of the $45°$ set-aquare, or by placing the leg of the compasses or dividers on O and swinging round the radii OP and OQ to the XY line at P_1 and Q_1.

Vertical lines are then drawn from these new points P_1 and Q_1 to meet the projectors from the horizontal edges of the elevation. In this way the end elevation is completed.

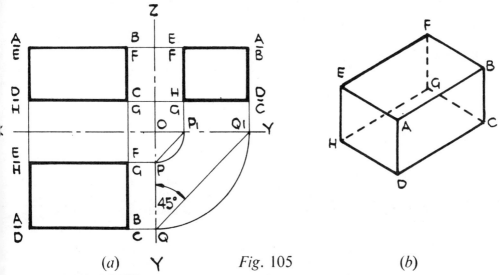

(a)   Y   Fig. 105   (b)

In practice, after a time, the ground and vertical hinge lines are omitted. The elevation is set out, the plan drawn in projection a convenient distance beneath it, and the end elevation drawn in projection with the front elevation the same distance from it as the plan by using the bottom right-hand corner $\frac{C}{G}$ as the centre of the arc.

It is advisable, however, to use the ground and vertical hinge lines as above until the relationship between the views is thoroughly understood.

Note that each corner of Fig. 105(b) is given a letter. In Fig. 105(a) the corners have been lettered to correspond. In each of the three views a corner is hidden behind the one shown. In order to make this clear, the corner nearer to the observer is represented by the letter above the one which represents the corner farther away.

In projection, when points or corners are indicated with letters or figures, this method should be adopted to avoid confusion, and great care should be taken to indicate the corresponding points in each view in the correct manner.

C. REPRESENTATION OF SURFACES

To Draw the Three Orthographic Views of an Equilateral Triangle as it Stands Perpendicular to the H.P. and V.P. (Fig. 106)

Suppose an equilateral triangle is cut from a piece of thin sheet metal. Fig. 106 shows the three views of the triangle when it stands at right angles to the H.P. and V.P. Note that the front elevation and plan are simply straight lines.

If each corner is lettered as in the end elevation, the corners in the other views can be located as shown. The two corners in line along the base in the front elevation, are indicated by $\frac{B}{C}$, which shows that B is nearer the observer as explained above.

You will realize that the height of the elevation is not the length of one side of the triangle, but the vertical height AD.

Draw these three views of the triangle as shown, assuming the length of side to be 40 mm. In this case you will need to begin with the end elevation.

To Draw Three Views of a Triangle when it is Inclined at $45°$ to the H.P. (Fig. 107)

It is first necessary to draw the true shape of the triangle ABC to obtain the vertical height AD. In this case the true shape is called an *auxiliary* (or *aiding*) *view*. Then draw the elevation equal to AD and inclined at the given angle to the H.P. Produce the horizontal and vertical projectors from A to D. Draw a ground line a suitable distance below D and a hinge line the same distance to the right of D. Mark the corner C on the appropriate projector the same distance below the ground line as D is above it, then points D and B equal to CD and CB on the auxiliary view. From D draw DA perpendicular to the projector, then join AB and AC.

To obtain the end elevation, produce the points CDB in the plan to the hinge line ZY, then project these points to the XY line at an angle of $45°$. From the XY line project the points vertically to meet their corresponding horizontal projectors from the elevation. The intersection of the appropriate projectors gives the points ABC in the end elevation, which are joined to complete this view.

Draw the three views of the triangle, if the vertical height is 45 mm.

To Draw Three Views of a Rectangle Inclined at $60°$ to the H.P. (Fig. 108)

Let a piece of thin sheet metal be cut out to form a rectangle and then placed as in Fig. 108. The three orthographic views are obtained in a way similar to the above.

First draw an auxiliary view (the true shape) and letter the corners A, B, C and D. This time the vertical height is the same as AD, so the elevation can be drawn that length at the given angle. Draw the XY and ZY lines as before, then drop vertical projectors from B and C in the elevation to give the width of the rectangle as in the plan view. Draw a horizontal line AD the same distance below XY as point C is above it, then

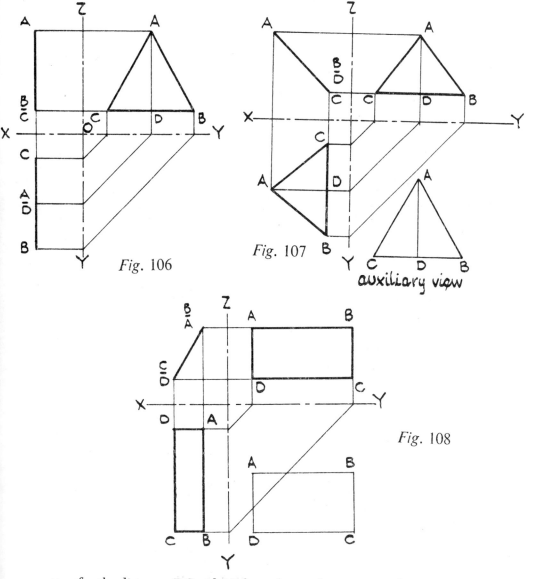

Fig. 106

Fig. 107 auxiliary view

Fig. 108

transfer the distances DC and AB from the auxiliary view to the corresponding projectors to complete the plan.

For the end elevation, draw the horizontal projectors from B and C in the elevation to give the height of the rectangle as seen in the end elevation. Draw a vertical line AD the same distance from ZY as AD is below XY. The lengths of AB and DC can either be measured direct along the horizontal projectors, or can be marked from the ZY line at 45° as shown. Join the points to complete the view.

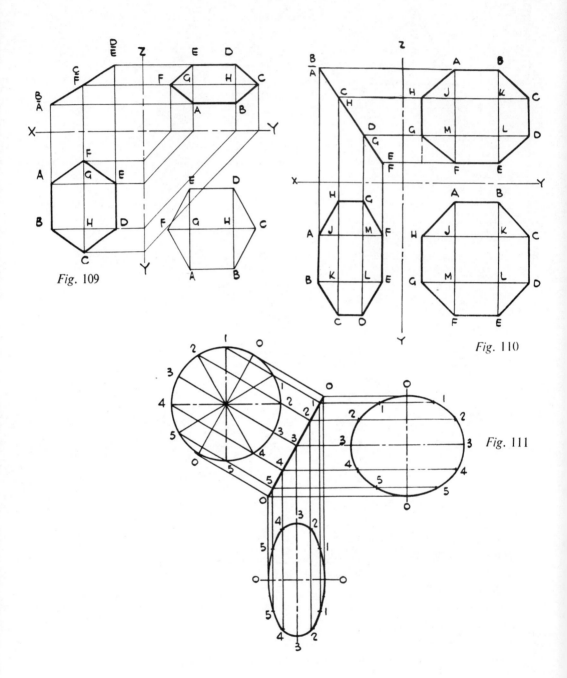

Fig. 109

Fig. 110

Fig. 111

To Draw Three Views of a Regular Hexagon when it is Inclined at 30° to the H.P. with one Edge Perpendicular to the V.P. (Fig. 109)

Draw the true shape and join the opposite corners. The elevation will be a line at 30° to the H.P. and equal to AE in length. Letter the true shape and elevation to correspond, including the point $\frac{C}{F}$. Draw the XY and ZY lines as before.

To obtain the plan, drop projectors from the three points in the elevation and mark F the same distance below XY as $\frac{B}{A}$ is above it. Mark FG, FH and FC along this projector of the centre line, then draw horizontal lines through the points G and H to give the corners A, B, C, D and E, which when joined complete the plan.

To obtain the end elevation, draw horizontal projectors from B, C and D in the elevation to give the height as seen in the end view. Draw projectors from F, G, H and C in the plan to the ZY line, then at 45° to the XY line, and finally vertically upwards to meet the corresponding projectors already drawn. Join the points so formed to complete the end elevation.

Draw the three views of the hexagon as above, if the length of each side is 25 mm.

If you are able to complete the above exercise, there is no reason why you should not be able to carry out the exercise of Fig. 110. This shows a regular octagon—assume the length of each side to be 20 mm—standing with one edge on the horizontal plane and inclined to that plane at 60°. The exercise is worked out fully for reference, if necessary.

To Draw the Three Views of a Thin Circular Disc Inclined at 60° to the H.P. with its Horizontal Diameter Perpendicular to the V.P. (Fig. 111)

First draw the elevation, a line at 60° to the H.P. equal in length to the diameter of the circle.

Unlike the other shapes dealt with so far, there are no corners for reference, so points are introduced round the circle. It is advisable to introduce these points evenly round the shape, so proceed as follows.

Draw perpendiculars from each end of the elevation, and then a centre line OO parallel to the elevation. Draw a true shape (a circle, and the auxiliary view in this case) between the projectors with OO as diameter, and divide the circle into 12 equal parts. (This can be done quickly with the compasses, using the radius of the circle, or it can be done with the aid of the 60° set-square.) Number the points as shown—or letters can still be used if preferred—and then draw parallel projectors perpendicular to OO through the corresponding points to meet the elevation. Number the points in the elevation to agree with those round the true shape.

For the plan, a horizontal line OO corresponding to OO in the elevation is drawn a suitable distance below it. (Notice that at this stage we have omitted the ground line, while all the projectors are drawn as they should be in future, namely as thin continuous lines.)

Vertical projectors are produced from the points in the elevation to cut the centre line OO in the plan. To obtain the required shape, the perpendicular distances of the points, 1, 2, 3 etc. from the line OO in the circle are now marked on the corresponding projectors from OO in the plan. This will give 1, 2, 3 etc. in the plan and when these points are joined up with a smooth curve the view will be completed.

The end elevation is obtained in a similar way. A vertical centre line OO is drawn the same distance from the upper O in the elevation as the centre line of the plan is below the base line. Horizontal projectors are produced from each point in the elevation to cross OO, and once again the perpendicular distance of each point from OO in the true shape is marked on the appropriate projector from OO in the end elevation. Finally, these points are joined with a smooth curve to give another ellipse, the shape of the disc as seen in the end elevation.

When the above method has been learnt, draw the three views of a circle of 60 mm diameter and standing at 30° to the H.P., without reference to Fig. 111 if possible.

D. REPRESENTATION OF PRISMS

In this chapter all solids are regular and resting parallel to the V.P.

The drawing of prisms follows on easily from the drawing of surfaces. Two surfaces identical in outline often have to be drawn for the ends, though some edges will be hidden and have to be shown as broken lines. The opposite corners are then joined with parallel lines to indicate the faces.

To Draw Three Views of a Rectangular Prism with One Edge Resting on the H.P. and One Face at 30° to the H.P. (Fig. 112(a))

For the elevation, begin with B on XY and draw BF at 30°, to the given length. Complete the rectangle ABFE. Drop projectors from each corner for the plan and make AD equal to the width of the prism. The end view

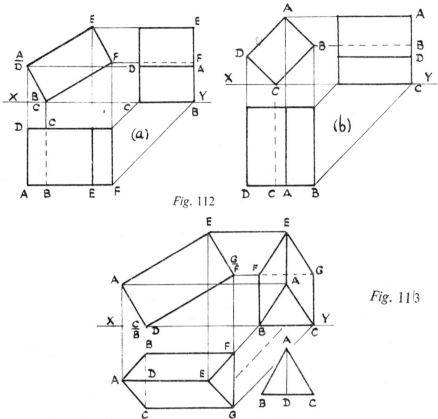

Fig. 112

Fig. 113

is completed by means of the horizontal and oblique projectors as shown. Note the broken lines for hidden edges. Let AB = 25 mm, AE = 50 mm, and AD = 35 mm.

Fig. 112(b) shows the same prism resting on one edge, but now with two faces at 45° to the H.P. Begin with the elevation and project in each direction as shown.

To Draw Three Views of a Triangular Prism Inclined at 30° to the H.P. (Fig. 113)

First draw the auxiliary view, then the line CG inclined at 30°, equal in length to that of the prism. On CG draw the rectangle ACGE, making CA equal to the vertical height AD in the auxiliary view. To complete the other views, drop vertical projectors from each point in the elevation to outline the plan, and horizontal projectors from the same points to help in forming the end elevation. Insert hidden edges.

Draw the above prism if BC = 35 mm and AE = 65 mm.

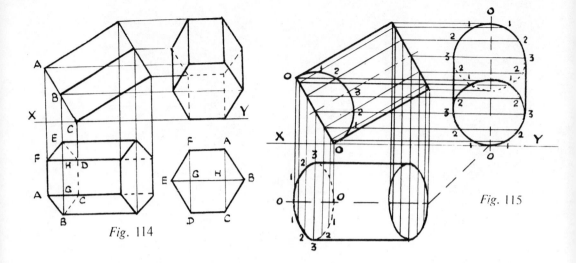

Fig. 114

Fig. 115

To Draw a Hexagonal Prism, with an Edge of One End Resting on the H.P. and the End Inclined at 60° to It (Fig. 114)

Copy the views as an exercise, studying the lettering carefully. Let the ends be of 25 mm length of side, and the prism 50 mm long. Refer to Fig. 114 only if necessary.

To Draw the Three Views of a Cylinder Resting on the H.P. with its Side Inclined at 30° to the H.P. and its Axis Parallel to the V.P. (Fig. 115)

This is a simple extension of Fig. 111.

First draw the elevation, a rectangle inclined at 30° to the H.P. of the same length as the cylinder and width equal to its diameter.

When drawing the disc of Fig. 111 you may have noticed that it was unnecessary to draw a separate auxiliary view. In Fig. 115 only a semicircle has been used, while to save time it has been drawn on the elevation OO, and then divided into six equal parts. Through the points on the semicircle draw projectors parallel to the axis of the cylinder. The rest of the construction follows Fig. 111, except that each view is made up of two ellipses instead of one, joined by parallel lines.

Fig. 116

E. REPRESENTATION OF PYRAMIDS

The additional work involved when drawing views of pyramids is the location of the apex in each view. The corners of the base are then joined to the apex, remembering that hidden edges must be indicated by broken lines as before.

If the vertical height of the pyramid is given, the construction is an easy matter, but when the length of the slant edge (usually called the *slant height*) is given, the vertical height has to be worked out.

To Draw the Three Views of an Equilateral Triangular Pyramid Standing on the H.P. (Fig. 116)

Suppose the slant height is given and not the vertical height. By studying the figure carefully, it will be realized that none of the edges DA, DB and DC in the elevation are true lengths of the slant edges of the pyramid, since the plan shows that each of these edges is inclined towards the apex.

If this is difficult to visualize, consider DC in the plan. Place the point of a pencil on C and incline the pencil away from you in the direction of D. If the point is held at eye level, it will be obvious that the pencil appears foreshortened. Carry out this simple demonstration for the lines DA and DB in the plan.

One method of obtaining the vertical height when the slant height is given is shown in Fig. 116. First draw the plan of the pyramid, locating D by bisecting the angles of the base triangle ABC. Draw a horizontal line through D; this will be parallel to the ground line, in this case the H.P. With D as centre and DC as radius, mark E on this line, then project the point to the H.P. Project D to the H.P. at C and erect a perpendicular of any length from C. With E on the H.P. as centre and radius equal to the slant height, mark D. DC will be the vertical height of the pyramid, and also, in this case, the elevation of that edge.

What has really been done is to turn the edge DC in the plan round, so that it is parallel to the V.P. and we are then able to see its true length.

It is advisable to follow the method carefully until the principle is understood, for this problem will be met again in a later chapter.

To complete the three views, project the positions of all corners to the H.P., and join A and B also to D. Produce BB for an auxiliary hinge line and draw horizontal projectors from D and C for F and C_1, then project B, F and C_1 to the H.P. for the end elevation as before. Erect the axis at F and draw a horizontal projector from D for the apex in the end elevation. Join B and C to D.

Notice particularly the difference between the two elevations.

Draw the three views of an equilateral triangular pyramid as in Fig. 116, if the edge of the base is 40 mm and the slant height is 65 mm.

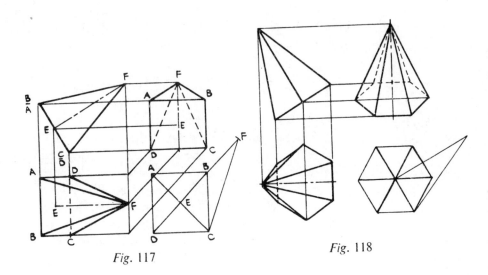

Fig. 117

Fig. 118

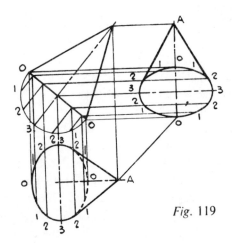

Fig. 119

18

To Draw the Three Views of a Square Pyramid Standing on the H.P. with its Base Inclined at 60° (Fig. 117)

Let the axis of the pyramid be parallel to the V.P.

Again, if the vertical height is given, the construction is straightforward. The auxiliary plan shows another method of finding the vertical height if the slant height only is given.

First draw the auxiliary plan as shown and mark the centre E. Draw a line from F perpendicular to EC. In the case of a square base, this can be quickly done by producing EB. From C mark CF equal to the given slant height. EF will then be the vertical height of the pyramid. Though the method may seem easier, the underlying principle is the same as that used in Fig. 116.

The elevation is then drawn, BC being the length of the side of the base and inclined at the given angle to the H.P. Mark E, the mid point of BC, and draw the axis from E perpendicular to the base and equal to EF in the auxiliary view.

The rest of the construction should present no difficulty.

Draw three views of a square pyramid in the position shown in Fig. 117, if the edge of the base is 40 mm and the slant height is 60 mm.

To Draw Three Views of a Hexagonal Pyramid, Resting on the H.P., with the Base Inclined at 30°, the Axis Being Parallel to the Vertical Plane (Fig. 118)

Attempt this construction, without reference to Fig. 118, given that the distance across the opposite corners of the base is 45 mm and the slant height is 60 mm.

To Draw Three Views of a Cone Resting on the H.P. with its Base Inclined at 45°, its Axis Being Parallel to the V.P. (Fig. 119)

This problem follows naturally from Fig. 111. The elevation of the cone is easy to draw when either the vertical height or the slant height is given. Notice that the semicircle representing a half true shape of the base is drawn below OO in the elevation, to avoid cutting across the slant sides in that view.

Draw three views of a cone inclined at 45° to the H.P., but in the opposite direction to that of Fig. 119, its axis being parallel to the V.P. Assume the base to be 60 mm and the vertical height 70 mm.

CHAPTER THREE

Isometric Drawing

Pictorial drawing was mentioned on page 6 as being another method of showing solids in geometrical and engineering work. Two forms of pictorial drawing are isometric drawing and oblique drawing, and in this chapter we will deal with the former.

Isometric drawing is given the name 'pictorial' because it produces a 'picture' view of the object. Refer back to Fig. 105(b). There the rectangular block is drawn in depth, so that it appears to stand out from the paper.

You will appreciate that it is much easier to visualize the block from this drawing—which is an Isometric one—than it is to understand the orthographic projections of the same block in Fig. 105(a).

This is one advantage of isometric drawing. Another is that the one view can often be drawn more quickly than the three orthographic views. Furthermore, length, breadth and height are clearly shown in the one picture.

On the other hand, the angles in some objects may appear distorted when drawn isometrically, and occasionally it is difficult to dimension the object properly.

Isometric really means 'equal measure', and this type of drawing—unlike oblique drawing—is measured out evenly on each side of a vertical line. This fact often prevents confusion with oblique drawing.

Isometric drawing is said to be based on three axes—or lines—at 120° to one another. This means that the lines are drawn as in Fig. 120, one line being vertical and the two others going in opposite directions at 30° to the horizontal. Along these three lines, lightly drawn with the aid of the 60° set-square, the length, breadth and height of the object are measured, and the points so obtained are joined up with lines parallel to the axes, except for angles other than right angles, to complete the view.

One corner of the object is always nearest to the observer in this method, and if the three axes are set out first, and regarded as forming the nearest corner, the object can be built up on this. The method will be found fascinating and fairly easy.

(Isometric *projection* should not be confused with isometric *drawing*. In the former method, many of the lines are foreshortened and an *isometric scale* is often used, but since this is not required to the standards set here, it will be omitted.)

Hidden edges are not usually shown in isometric drawing unless required for some special reason, and they should not be indicated unless asked for.

It is important to remember that all measurements of the object have to be made along the axes or lines parallel to them. Also, angles cannot be set out as they really are in the object. They can only be drawn by taking measurements along the axes and joining the points so marked.

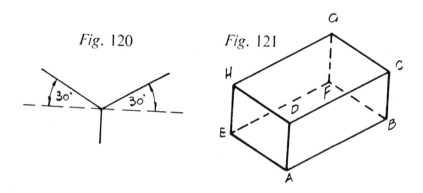

Fig. 120   Fig. 121

To Make an Isometric Drawing of a Rectangular Block (Fig. 121)

Many students in making an isometric drawing find it easier to begin with the vertical axis of Fig. 120 above the horizontal line instead of below it. This is quite in order, if preferred, and means that the bottom corner EAB of Fig. 121 will be drawn first. The exercise will be described in this way.

Begin with the three axes as above and mark off on these the length AB, the width AE and the height AD. Erect vertical lines at B and E. Draw lines from D parallel to AB and AE for corners C and H. Draw a line from H parallel to DC, and another from C parallel to DH to complete the figure at G.

Unless you are instructed otherwise, the object can also be drawn with the edge EH nearest the observer.

To Set Out Angles in Isometric Drawing (Figs. 122(a) and (b))

Fig. 122(a) shows an elevation and plan of a tapered block. As we cannot set out the taper directly in isometric drawing, we adopt the following procedure.

Produce the parallel edges of the block in each view so that a rectangle is formed. This is known as ' putting the object in a crate '. The phantom

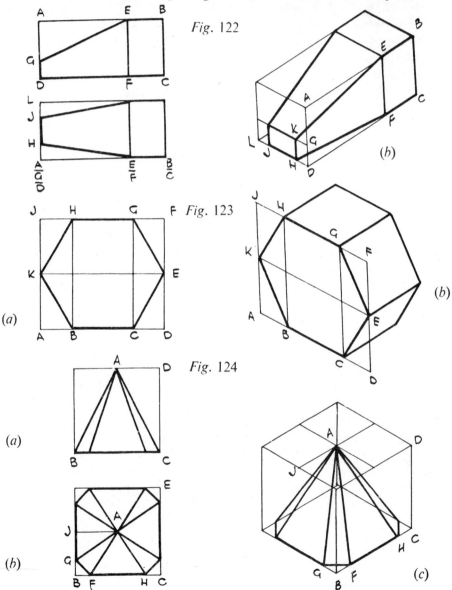

Fig. 122

Fig. 123

Fig. 124

crate is then drawn isometrically, and all measurements are made along the edges of this outline.

For instance, to obtain the line FH in the isometric view, mark off the distance CF in the elevation along CD on the crate, then mark off the distance DH in the plan from D and along DL on the crate. Join FH.

To draw the line EK, mark E along BA as above. Mark the height DG in the elevation on DA on the crate, and through G draw a line at 30° (that is, parallel to DL). Where this line intersects a vertical line drawn from H on the crate will give point K. Join EK. Locate the other points in a similar way.

If two or more points lie along the same edge of a crate, it is advisable to locate the points from the same corner to avoid confusion.

Make an isometric drawing of the tapered block in Fig. 122(a), if AB = 70 mm, DL = 40 mm, DH = LJ = 10 mm and DG = BE = 20 mm.

To Make an Isometric Drawing of a Hexagonal Prism (Figs. 123(a) and (b))

An elevation of the prism is shown at (a). Assume the prism to be 25 mm long.

The method, again, is to draw a crate round the object. This crate ADFJ is then drawn isometrically. The distances DC and DB in the elevation are marked from D along DA on the crate, and the distance DE in the elevation is marked from D along DF on the crate. Lines are drawn vertically from B and C, then one at 30° through E. These lines will give the points H, G and K on the crate, and when these are joined as in the elevation the end of the prism is finished.

To complete the drawing, extend lines from C, E, G and H at 30°, make those lines 25 mm in length (the length of the prism), then join the end points so marked to give the visible outline of the farther end.

Once more it must be emphasized that isometric drawing can be done quickly with the aid of the 60° set-square.

To Make an Isometric Drawing of a Square Pyramid (Figs. 124(a), (b) and (c))

Assume the pyramid to be chamfered or bevelled at the corners as shown in the elevation and plan. Draw a crate around these views, then the full isometric cube as at (c). Notice how the apex A is located. As the figure is a square pyramid A is centrally placed in the upper isometric square. To obtain the position of the chamfer, mark BF and BG on the cube equal to BF and BG on the plan. Join F and G to A. Repeat this method for the other visible chamfers.

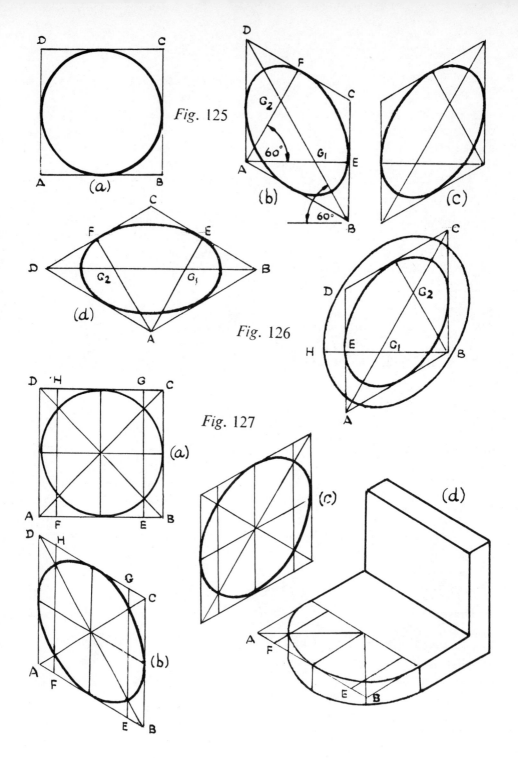

The Isometric Circle

Often it is required to draw circles or curved shapes isometrically. Three methods will be shown.

A. The American Method (Fig. 125(a), (b), (c) and (d))

This method is the easiest because the ellipse so formed can be drawn with the compasses. Fig. 125(a) shows the required circle inscribed in a square. A square of the same size is then drawn isometrically as at (b) or (c). The side BC is bisected at E (this is done quickly by means of a horizontal line), and the side CD by the line AF (drawn at 60°). The diagonal BD is also drawn at 60° to the horizontal, and the intersection of this diagonal with AE and AF will give G_1 and G_2. With centre G_1 and radius G_1E draw an arc, and with centre G_2 and the same radius draw a second arc. Then with centres A and C and radius AF draw joining arcs to complete the shape. Figs 125(c) and (d) indicate the methods for the circle in the other positions.

Fig. 126 shows a quick method for drawing concentric circles isometrically. Draw the inner 'circle' as before. Produce BE and make EH equal to the difference in radii. Then with the same centres, but with G_1H and BH as radii construct the outer 'circle'. If it is required to draw the circles more accurately, for they should not appear as concentric in the isometric view, a second isometric square should be drawn and the outer circle constructed from it.

The American method is a quick and satisfactory method for beginners. But it is not a true isometric reproduction, and should not be employed where accuracy is essential, nor in G.C.E. examinations.

B. The Diagonal Method (Figs. 127(a), (b), (c) and (d))

Draw the required circle, enclose it in a square and divide the square into four. Construct the diagonals and erect vertical ordinates EG and FH where these cut the circle as shown. Construct the isometric square, dividing it also into four equal parts, then mark off along the base the distances BE and BF. Erect vertical ordinates through points E and F, and where these ordinates cross the diagonals will give four other points on the curve. Join the points carefully as shown. If you can follow this construction you should find it easy to draw also the horizontal isometric circle, and to make a full size copy of the bracket at (d).

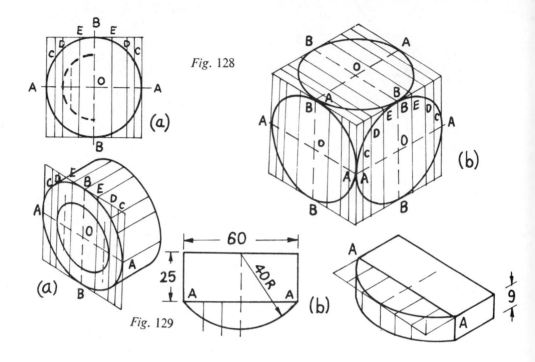

Fig. 128

Fig. 129

C. *The Ordinate Method* (Figs. 128(a) and (b))

This is the most accurate method. The circle and square are first drawn as before. The horizontal and vertical diameters AA and BB are drawn, and AA is divided into a suitable number of parts. These need not be evenly spaced, for the end division should be sub-divided as shown in (a) for greater accuracy when drawing the curve. Draw ordinates parallel to BB through the points on AA.

An isometric square equal to the diameter of the circle is drawn in a vertical or horizontal position as required and divided into four parts by AA and BB. The positions of the ordinates along AA from O in the elevation is marked from O along AA on the isometric square. The ordinates are then drawn parallel to BB, and lettered to correspond with their respective ordinates in the elevation.

The length of each ordinate above and below AA is transferred to the isometric square, and the points C, D, E etc. are joined to form a smooth curve both above and below AA. In using this method it is preferable to use a few ordinates accurately than too many inaccurately.

As an exercise draw a circle of 50 mm diameter isometrically, using the three methods described and without referring to the figures, if possible.

To Draw a Cylinder Isometrically (Fig. 129)

Draw one face of the cylinder as above. From A and B and the end of each ordinate C, D, E, draw projectors at 30° to the horizontal. Mark off the length of the cylinder, say 75 mm, on each of these projectors, and join the resulting points with a smooth curve.

Fig. 129(b) shows how to set out any curve by the ordinate method. It should be copied full size.

EXERCISE 1

1. A hexagonal prism has a length of 50 mm while each side of its end is 20 mm long. Make an isometric drawing of the prism resting in any suitable position.
2. Make an isometric drawing of an octagonal prism resting with one end on the H.P. Assume the prism is 50 mm across and 60 mm long.
3. Fig. 130 is a pictorial view of a bracket. Draw the bracket full size, not as shown but isometrically as seen from arrow 'A'.
4. Fig. 131 shows the elevation of a ridge tile. Make an isometric drawing of the tile, assuming it to be 80 mm long.
5. A piece of tube 60 mm long has an external diameter of 50 mm and an internal diameter of 30 mm. Make an isometric drawing of the tube, showing your method clearly.
6. A right cone has a base of 70 mm diameter and a vertical height of 75 mm. Make an isometric drawing of the cone when it stands with its base on the H.P.
7. Fig 132 shows the elevation and plan of a cast iron bracket with dimensions as shown. Draw the bracket isometrically as seen in the direction of arrow 'Y'.

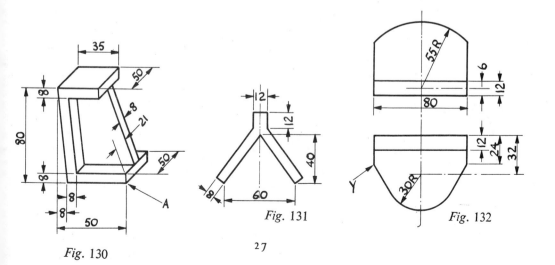

Fig. 130   *Fig.* 131   *Fig.* 132

8. Two simplified views of a casting are shown in Fig. 133. If the dimensions are given, draw an isometric view of the casting, showing all hidden lines.

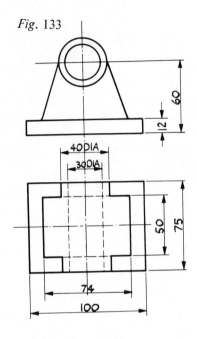

Fig. 133

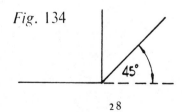

Fig. 134

CHAPTER FOUR

Oblique Drawing

Oblique drawing is the other method of pictorial drawing used in geometrical and engineering work. It also is based on three axes, but two of these are at right angles to each other and the third is at 45°, as shown in Fig. 134. Sometimes this third axis is drawn at 30° instead of at 45°, but it is advisable to use an axis at 45° to distinguish between this and the isometric method, unless instructed otherwise.

An object drawn on these axes has one face parallel to the V.P., and this face is therefore a true elevation. This fact makes oblique drawing easier than isometric drawing in most cases. It is best to draw the longest face in elevation, unless another face of the object is of an irregular or more complicated outline, when that face should be constructed on the two angles at right angles.

All measurements along the 45° axis must be made *half size* to the rest of the drawing to avoid distortion. This means that if a line 60 mm long is being drawn to a scale of half size, instead of making the line 30 mm long (normal half size), it should be made 15 mm along the 45° axis or lines parallel to it.

If the reason for these half-size measurements cannot be understood, draw a 50 mm cube obliquely with the 45° axis 50 mm long, then make another drawing with the same axis only 25 mm long. It will be seen at once which drawing looks more like a cube.

The oblique drawing is built up on these axes by transferring measurements from the object along them or on lines parallel to them. All angles along the 45° axis can only be drawn by taking measurements along it—remembering that the distances must be half size compared with the rest of the drawing—then measuring along the other axes and joining the points.

As in isometric drawing, hidden edges are not usually shown unless for some special reason, and they should not be drawn unless requested.

If this method of drawing is likely to be confused with the previous one, it should be remembered that in isometric drawing *one corner* of the object is nearest the observer, while in oblique drawing *one whole side* is nearest the observer, as can be seen by referring to Fig. 130.

Fig. 135 shows an oblique drawing of the same rectangular block as in Fig. 121. Compare the two drawings, noting particularly that the lines BF and CG are drawn half the length of these lines in the isometric drawing.

Study Fig. 136. This is an oblique drawing of the tapered block of Fig. 122(b). The block has been drawn from the end direction because it will better emphasize the taper. Notice that the block has been drawn with the aid of a crate in the same way as in isometric drawing, but in this case the crate is drawn obliquely.

Most objects to be drawn obliquely require to be 'crated' in this way, and it must be emphasized that all angles cannot be drawn as in the original

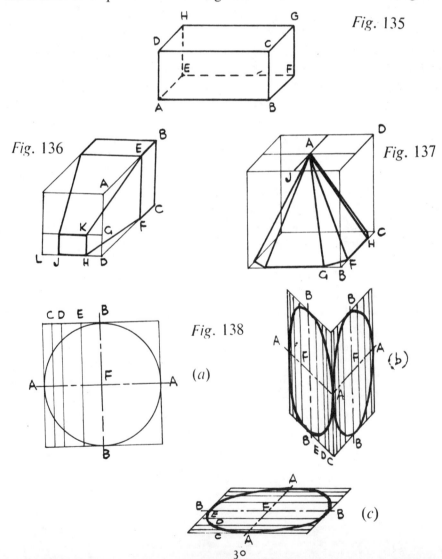

view, but must be measured by means of distances along the axes as in isometric drawing. The plan and elevation of the block would, of course, have to be drawn as in Fig. 122(a) before the oblique view can be attempted.

Fig. 137 shows an oblique drawing of the square pyramid of Fig. 124. Draw the elevation and plan first as in Figs. 124(a) and (b). Construct an oblique crate to the required size and mark off the distances of the chamfer along the horizontal and 45° axes, remembering that the latter must be one half of those shown in the orthographic views. Locate the apex as shown, then join up all points to complete the view.

To Draw a Circle Obliquely When It Stands in the V.P. (Figs. 138(a) and (b))

Draw the circle, enclose it in a crate and divide AF into a suitable number of parts. You may have already realized that it is necessary to divide only one half of AA, in the case of the circle, to save time.

Draw an oblique square by making BB equal to BB in the true shape and AA equal to AF. The position of the ordinates from BB is then marked along AA in the oblique view, but as this is the 45° axis the distance of each ordinate from BB is made only half the horizontal distance from BB in the true shape. The length of each ordinate from AA is then plotted as before and the points joined to give the oblique shape. The circle can, of course, be drawn in either direction, as shown.

To Draw a Circle Obliquely When it Rests on the H.P. (Fig. 138(c))

An oblique square is drawn but in this case BB is drawn horizontally and AA at 45°. The distance of the corresponding ordinates from BB in the oblique view is made one half the distance from BB in the true shape. The ordinates are drawn parallel to BB, then the lengths are marked off and the points joined to form a smooth curve as before.

To Draw a Cylinder Obliquely

Unless you are otherwise instructed, draw the cylinder with its circular end parallel to the V.P. Draw tangents to the circle at 45° making them half the length of the cylinder. Project the axis of the cylinder also at 45° and half the true length and describe an arc to represent the other end.

If the cylinder has to be drawn in any of the other positions, draw the complete circular end first, extend projectors from points on this oblique circle making them the length of the cylinder, then join up the required points to show the outline of the other end similar to the method of Fig. 129.

EXERCISE 2

1. Make an oblique drawing of the ridge tile shown in Fig. 131.
2. Given that the dimensions of a standard housebrick are 75 × 99 × 228 mm, make an oblique drawing of a brick one third full size.
3. A cone has a base of 60 mm diameter and a vertical height of 65 mm. Draw an oblique view of the cone as it stands with its base on the H.P.
4. A cylinder, 50 mm in diameter and 65 mm in length, is pierced along its axis by a centrally placed square hole the diagonal of which is 30 mm. Draw an oblique view of the cylinder as it stands with one end on the H.P.
5. Two views of a crank arm, blank at one end, are given in Fig. 139. Make an oblique drawing of this.
6. Fig. 140 shows the elevation of a pyramid whose base is a regular octagon. Draw an oblique view of the frustum (the lower part).
7. Fig. 141 shows a geometrical model consisting of a frustum of a cone standing on a chamfered square base. Draw an oblique view of the model.
8. Two views of a casting are given in Fig. 142. Make an oblique drawing of this casting.

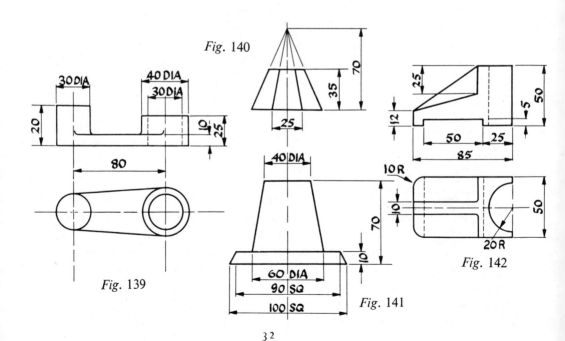

Fig. 140

Fig. 139

Fig. 141

Fig. 142

CHAPTER FIVE

Sections

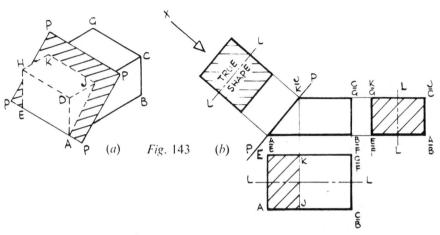

(a)   Fig. 143   (b)

A. SECTIONS OF SOLIDS

So far only outside views have been dealt with. Sometimes in geometrical drawing, and often in technical drawing, it is necessary to show how an object looks when it is cut through in a certain way, in order to give further information about it.

Fig. 143(a) is an isometric view of the rectangular block previously used. Imagine the block to be cut right through with a large blade (known as a *cutting plane*) PP, and the part to the left, shown in broken lines, removed.

An elevation of the block would be as shown at (b), with the cutting plane indicated by the line PP. Fig. 108 showed the method of producing the plan and the end elevation of such a shape, but it should be clear that none of these views gives the true shape of the section.

The true width is shown by the line AE in the plan, but the length AJ is not true. The true width is also shown by AE in the end elevation, but the length AJ, again, is not a true length.

The only view to give a true shape of the section is one at right angles to the cutting plane, that is, looking along the arrow 'X'.

Therefore, to draw the true shape it is necessary to project at right angles to the surface PP with a centre-line LL in Fig. 143(b) and build up the true shape on this centre-line by marking on it the widths taken from the centre line on the plan or on the end elevation.

33

It is usual to show the true shape adjacent to the cutting plane, as in the figure, for this often saves time, but it is not essential to do so.

The rules for drawing the true shape of any section are:
1. Draw projectors at right angles to the cutting plane from the edges of the cut surface, or from suitable points on it.
2. Draw a centre-line parallel to the cutting plane, and therefore at right angles to the projectors.
3. Transfer true widths of the surface from the centre-line on the plan or end elevation to the appropriate projectors, working from the centre-line on the true shape.

The cut surface (or section) is indicated by evenly-spaced parallel lines. These are usually drawn at 45° to the horizontal, unless the section is at such an angle that the section lines would be nearly parallel to the sides, as in the true shape of Fig. 143(b), when some other angle is chosen.

These lines are called *hatched* lines, while the procedure is known as *hatching*. The lines should be suitably spaced according to the area to be covered. They should always be drawn more lightly than the bold outline of the shape, and should exactly touch it for neatness, not falling short of the outline nor crossing it.

To Draw the True Shape of a Triangular Prism Cut by a Plane PP (Fig. 144)

Draw first the three views as in Fig. 107. Draw the projectors AA and BC perpendicular to the cutting plane. Draw the centre-line AD parallel to PP. Mark off the distances DB and DC from D, taken from the plan or end elevation. Join B and C to A and hatch all the sectioned surfaces.

Draw the four views of the prism, assuming it to have a length of side of 40 mm, an overall length of 55 mm, and the cutting plane to be at 45°.

To Draw the Three Views and True Shape of a Hexagonal Prism Resting with One End on the H.P. and Cut by a Plane PP (Fig. 145)

You should be able to draw the three views without referring to Fig. 110. The true shape is obtained by following the rules given above. Project perpendicularly to PP from points A, B, C and D in the elevation. Draw AD parallel to PP, and mark off the distances from AD on the true shape of B, C, E, F, (taken from the plan or end elevation) along their respective projectors. Join the points and hatch the sectioned surfaces.

Copy Fig. 145, assuming the prism to have a length of side of 20 mm and a vertical height of 50 mm. Assume PP to be at 30° to the horizontal.

To Draw the True Shape of a Cylinder Resting with One End on the H.P. and cut by a Plane PP (Fig. 146)

The elevation and end elevation are drawn as in Fig. 111, projectors being used and carefully numbered. From the points O, 1, 2, 3, on PP draw further projectors at right angles. Draw OO parallel to PP, and transfer the distances of the points from OO in the end elevation from OO on the true shape along the respective projectors. Join up these points with a smooth curve. The end elevation has not been hatched so that the method is clearer.

The method is exactly the same when the cylinder is resting with its axis parallel to the H.P.

Copy Fig. 146 if the cylinder has a diameter of 50 mm and PP is at 45° to the H.P. (In the figures below, the original lengths of the solids have been omitted for clarity.)

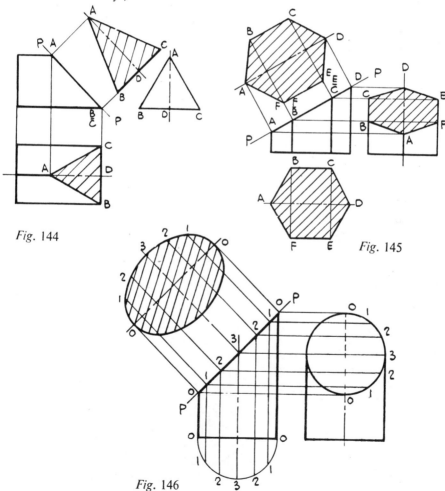

Fig. 144

Fig. 145

Fig. 146

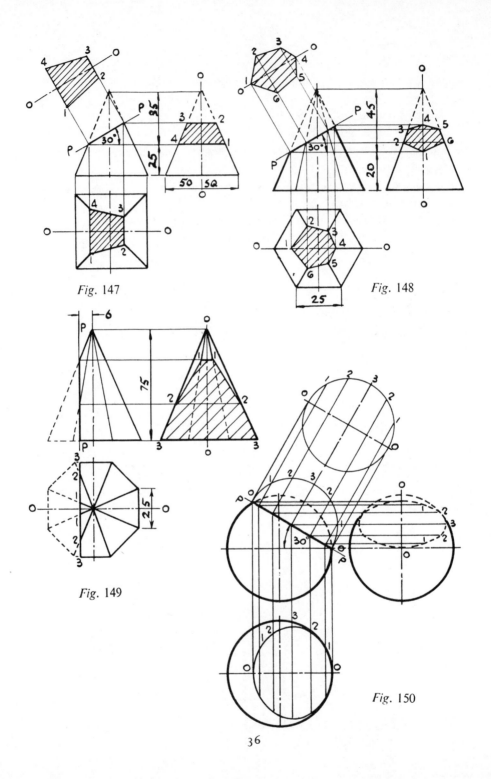

Fig. 147

Fig. 148

Fig. 149

Fig. 150

To Draw Three Views and a True Shape of a Square Pyramid Cut by a Plane PP (Fig. 147)

Draw the three views of the complete pyramid in light outline. Introduce the cutting plane PP in the elevation. Project to the plan and end elevation from the points where PP intersects the edges of the pyramid. These projectors will give 1, 2, 3, and 4 in the other views, the points being joined as shown. The true shape is then obtained in the usual way. Line in the views as shown.

Draw Fig. 147 to the given dimensions without referring to the method, if possible.

To Draw Three Views and a True Shape of a Hexagonal Pyramid Cut by a Plane PP (Fig. 148)

Draw the three views as before, introduce PP and project to the respective edges in the plan and end elevation from the points where PP cuts the edges in the elevation. Project from these points also perpendicularly from PP for the true shape, draw OO parallel to PP and mark off the widths from the plan or end elevation. Line in correctly as before.

Draw Fig. 148 to the given dimensions.

To Draw Three Views of an Octagonal Pyramid Cut by a Plane at Right Angles to the V.P. and H.P. (Fig. 149)

This is a slightly different construction because the true shape is shown by the end elevation.

Draw the complete outline of the three views and the cutting plane PP. The plan of the section will be a straight line, as in the elevation. To obtain the true shape, project at right angles from PP, and transfer the widths of 1, 2, 3 from OO in the plan to 1, 2, 3 from OO in the end elevation. Join the points to obtain the true shape. Then line in.

Draw Fig. 149 to the given dimensions.

To Draw Three Views and a True Shape of a Sphere Cut by a Plane PP (Fig. 150)

Just as the view of a sphere from any direction is always a circle, so the true shape of any section of a sphere is a circle of diameter equal to the greatest width of the section.

Draw the three views in outline and the plane PP on the elevation. A semicircle is drawn on OO (where PP cuts the outline of the circle) and the rest of the construction is as Fig. 111. The true shape is a circle OO in diameter, but is usually drawn in projection as in the figure. The hatching has been omitted in all views so that the construction can be followed.

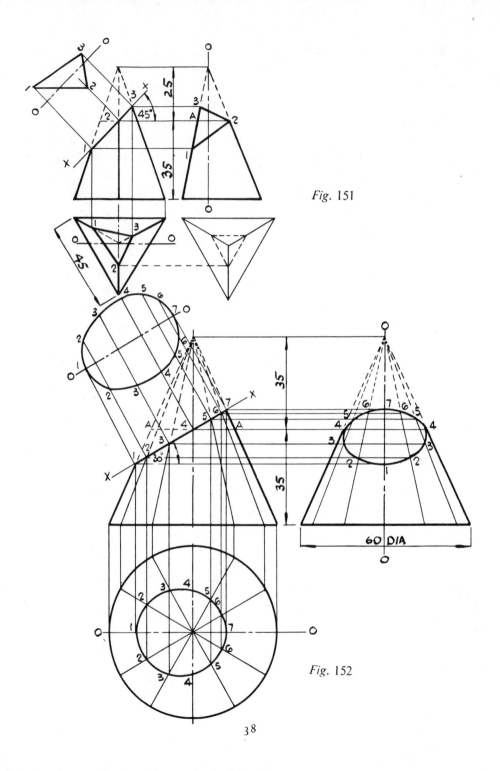

Fig. 151

Fig. 152

Draw the construction from memory, if possible, if the sphere has a diameter of 60 mm.

To Draw the Three Views and True Shape of a Triangular Pyramid cut by a Plane XX (Fig. 151)

This construction is a little more difficult. Draw the three views lightly in full and introduce the cutting plane in the elevation. Points 1 and 3 in the plan can be found as before by dropping projectors from 1 and 3 in the elevation. To find point 2 in the plan, project horizontally from 2 in the elevation to 2 in the end elavation. The perpendicular distance of 2 from OO in this view can now be transferred perpendicularly from OO in the plan.

The principle underlying this method is the introduction of a horizontal cutting plane through point 2 in the elevation. The line 2A will solve the view of the pyramid in the end elevation, while the corresponding plan view will be two concentric triangles as shown in the auxiliary plan, the broken lines indicating the view of the section.

Try to understand this principle, for we shall use it again in the examples on the cone.

To obtain the true shape, draw perpendicular projectors from 1, 2 and 3 and a centre line OO. Extend the vertical projectors from 1 and 3 to OO in the plan. Mark off the perpendicular distances OO1, OO2, OO3 in the plan, from OO on the respective projectors to form the true shape.

Draw the three views and the true shape to the dimensions given in Fig. 151.

To Draw Three Views and a True Shape of a Cone Cut by a Plane XX (Fig. 152)

More conic sections are shown on page 46. The three shapes we shall deal with are the *ellipse*, the *hyperbola* and the *parabola*.

There are two common methods for working out the sectional views of cones. In this case and the next, the simpler of the two is used.

Draw the three views lightly in full and also the cutting plane in the elevation. Again, as there are no edges for reference, light lines—traces—are introduced in the plan somewhat similar to the method of Fig. 119. The positions of these traces round the base—quickly made with the 60° set-square—are projected to the bases of the elevation and end elevation, and the points so marked are joined to the apex.

The intersection of XX with these traces will give 1, 2, 3 etc. on the cutting plane, and vertical projectors are dropped to the plan and horizontal projectors drawn to the end elevation from these points.

The importance of correct lettering and figuring in this work must be re-emphasized, and there should be little difficulty if the centre-line is used as a starting point in each view.

The projectors from 1 and 7 to the centre line OO will give the maximum length of the section in the plan, and the maximum height of the section in the end elevation. Next, where projector 2 intersects trace 2, projector 3 intersects trace 3, projector 5 trace 5 etc., in the plan and end elevation, will give points on the required shape. Point 4 cannot be found in this way. So it is imagined that the cone is cut horizontally through AA, as in Fig. 151. The plan of this section will be a circle of radius equal to 4A,

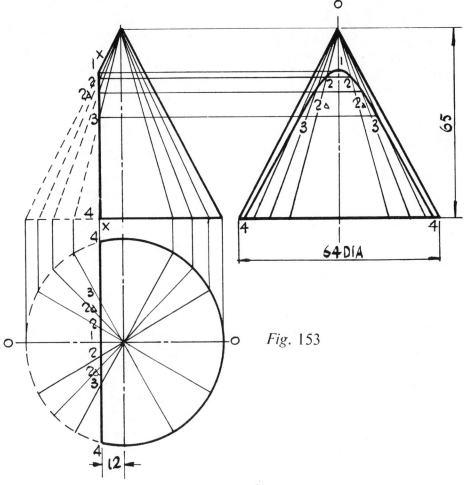

Fig. 153

so this radius is marked on the vertical centre line from the centre of the plan to give point 4 in that view.

The rest of the construction follows as before. Perpendicular projectors are drawn from 1, 2, 3 etc. on XX, a centre line OO is drawn parallel to XX, and the vertical distances from OO in the plan are marked off from OO on the true shape. Once again the hatching has been intentionally omitted. The lining in of views must be done neatly.

Carry out the exercise of Fig. 152, using the dimensions given.

To Draw the True Shape of a Cone Cut by a Plane XX Parallel to its Axis (Fig. 153)

Draw the three views lightly in full, and the cutting plane in the elevation and plan. Introduce traces on the plan as in the previous problem, mark the position of these traces on the bases of the other views, then join them to the apex in each case.

The intersection of XX with these traces will give the points 1, 2, 3 on the elevation. Draw horizontal projectors from these points to the end elevation. Where projector 1 intersects OO, projector 2 intersects trace 2, and projector 3 trace 3 will give the points on the required shape. Point 4 is marked direct from OO in the end elevation, using the perpendicular distance of 4 from OO in the plan.

If the points in the end elevation are too widely spaced, further traces can be introduced if necessary as at 2a in the plan, providing the positions of such traces are accurately marked in the other views. Horizontal projectors will then give the correct positions of these points on the true shape, as at 2a.

Carry out the exercise of Fig. 153, introducing additional traces as shown.

To Draw Three Views of a Cone Cut by a Plane XX Parallel to its Slant Side (Fig. 154)

We will use here the second method mentioned on page 39.

Draw the three views lightly in full and the cutting plane XX. Choose a number of horizontal cutting planes between 1 and 4 (where XX cuts the side and base of the cone) in the elevation. Such a plane is shown by the line 1A. The plan of this plane will be a circle of diameter 1A, and this circle is drawn lightly on the plan. The plan views of the other horizontal cutting planes are also drawn, using the horizontal distance from the axis to the *full outline* of the cone in each case.

Next, drop vertical projectors from points 1, 2 3 on XX to cut their respective circles. For example, the projector from 1 in the elevation will intersect OO for 1 in the plan, the projector from 2 will intersect the next circle at 2–2 in the plan, and from 3 will intersect the third circle at 3–3. The points 1, 2, 3, 4 are then joined with a smooth curve for the plan of the section.

The hatched plan gives the shape of the cone when it is cut by a horizontal plane through point 3 in the elevation. This may be of some help in following the principle which is employed.

To obtain the end elevation, draw horizontal projectors from 1, 2, 3 on XX to the end elevation and mark off the widths 2–2, 3–3, 4–4 from OO in the plan. Point 1 will, of course, lie on the axis.

This method can be used for any section of a cone, but the former rule applies. It is better to use a few cutting planes accurately than too many inaccurately.

Draw the three views of the cone in Fig. 154 and project a true shape of the section.

B. SECTIONS OF HOLLOW BODIES

Before the problem of sections is left, the case of solids pierced by holes must be considered.

Fig. 155(a) shows a model consisting of a cylinder attached to a hexagonal prism. The model is pierced right through with a horizontal circular hole, and the prism is pierced right through with a smaller vertical hole.

A section of the model through the plane AA looking in the direction of the arrows would give a view as at (b), a section through BB in the direction of the arrows as at (c), and through CC as at (d).

155(b) should be easy to follow, but notice in (c) that the hexagon is not hatched. This is because although the cutting plane is in contact with the end of the hexagonal prism, it does not cut it in any way. (d) shows the section where the centres of the two holes coincide, yet the horizontal hole is shown as a full circle because the outline of the hole will still be seen beyond the cutting plane.

Fig. 156 shows a hexagonal prism pierced by a square hole and cut by a plane DD. This section should also be easy to follow, and is formed by perpendicular projectors from DD in the elevation. Notice the horizontal line across the centre of the section. This is shown because it is a sectional *view*, and not just a true shape.

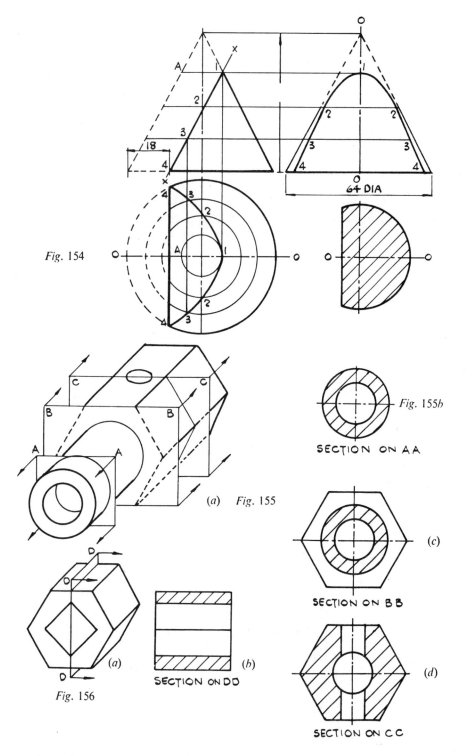

Fig. 154

Fig. 155b
SECTION ON AA

(a) Fig. 155

(c)
SECTION ON BB

(a)
Fig. 156

(b)
SECTION ON DD

(d)
SECTION ON CC

Fig. 157 shows a hexagonal prism pierced by a circular hole and cut by a plane EE. Copy this drawing full size as follows.

Draw the three views lightly in full and the cutting plane in the elevation. Project vertically from the points where EE cuts the prism and hole for the plan, and horizontally from these points for the end elevation. Hatch only the parts of the prism actually cut by the plane EE, and complete the views by showing the hidden edges.

The true shape is obtained in the same way as before. Project at right angles to EE from each point cut by the plane. Draw a line parallel to EE and mark from it the length of the prism on the projectors (or work from a centre line). Notice that hidden edges are not shown on the true shape. Hatch where necessary to complete the view.

To Draw Three Views and a True Shape of a Hexagonal Prism Pierced By a Circular Hole and Cut by a Plane YY (Fig. 158)

Draw the three views lightly in full. Choose suitable traces across the circle in the plan, and project these to YY. Notice that two of these traces pass through the edges B and C. This saves time, as four points can

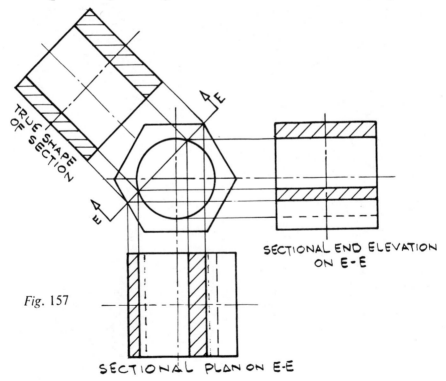

Fig. 157

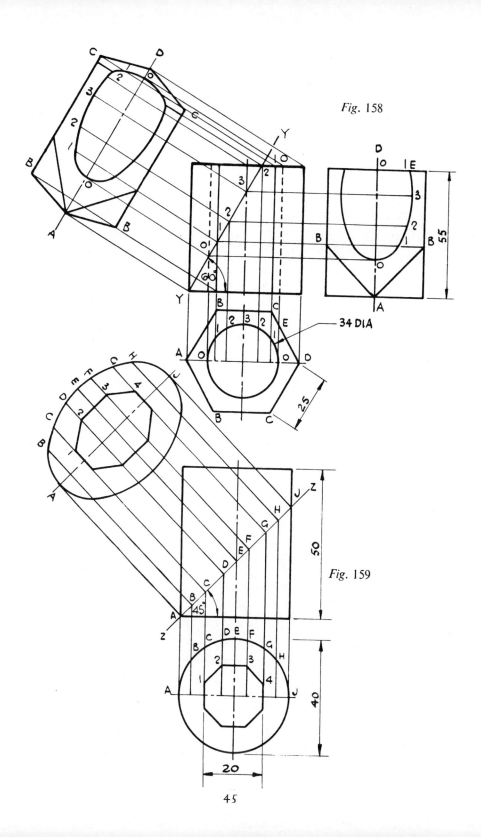

Fig. 158

Fig. 159

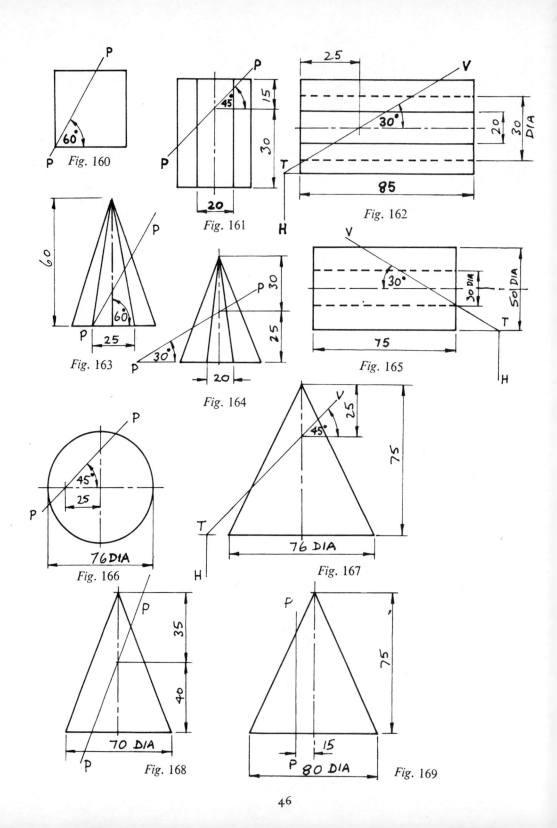

then be marked on the one projector. Project horizontally from all points on YY to the end elevation, and working from the centre line AD mark off the respective widths as before, locating point E in the end elevation in the same way. The hatching has been purposely omitted.

You should be able to work out the true shape without further instructions. Copy Fig. 158 to the dimensions given.

To Draw the True Shape of a Cylinder Pierced by an Octagonal Hole and Cut by a Plane ZZ (Fig. 159)

Try to draw this full size without reference to the method as shown in the figure. Add an end elevation and hatch where appropriate.

Often a cutting plane is shown as in Fig. 162. The edge-on view of a plane is a line and VT stands for *vertical trace*, which means the trace (or line) the inclined plane will make on the V.P., while HT stands for the trace the inclined plane makes on the H.P. Reference to Fig. 176 will make this clearer, where the bold line of $X_1 Y_1$ is the VT and the line crossing the horizontal plane is the HT of the inclined plane.

EXERCISE 3

1. Fig. 160 shows the elevation of a cube of 30 mm side cut by a plane PP. Draw a true shape of the section.
2. Draw an end elevation and a true shape of the hexangonal prism cut by a plane in Fig. 161.
3. Fig. 162 shows the elevation of an octagonal prism pierced by a circular hole and cut by a plane VTH. Draw a plan of the prism when it is so cut, and project a true shape.
4. A hexagonal pyramid is cut by a plane as shown in Fig. 163. Construct a plan, end elevation and true shape of the pyramid when it is so cut.
5. Draw the three views and true shape of the octagonal pyramid cut by a plane VTH in Fig. 164.
6. Fig. 165 is an elevation of a cylinder pierced by a circular hole and cut by a plane VTH. Draw the elevation and project the other two views and a complete true shape.
7. Draw the three views and true shape of the sphere cut by a plane in Fig. 166.
8. A right cone is cut by a plane VTH as shown in Fig. 167. Draw the elevation of the cone, and project a plan, end elevation and a true shape when it is so cut.
9. Fig. 168 shows a cone cut by a plane parallel to its slant side. Draw the three views and project a true shape.
10. The right cone in Fig. 169 is cut by a plane parallel to its axis as shown. Draw an end elevation of the cone.

CHAPTER SIX

Auxiliary Views

As mentioned in the previous chapter, a view of an object is often required other than the three usual orthographic views. The work on sections demonstrated this, where the true shape was projected at right angles to the cut surface. This was usually a view of the surface on a plane which was not parallel to the vertical, side vertical or horizontal planes.

Sometimes it is required to construct not the true shape of a section, but a view of an object as seen against a plane other than these three. Such a plane is known as an *auxiliary* (or *aiding*) plane, and the view so drawn is called an *auxiliary view*.

A. AUXILIARY ELEVATIONS

Study Fig. 170. This is an isometric view of the rectangular block lying parallel to the H.P. Instead of the usual S.V.P. an *auxiliary vertical plane* has been introduced at a certain angle marked ' θ ' (pronounced theeta), a letter of the Greek alphabet commonly used to indicate an angle.

Just as the end elevation is a view projected at right angles to the S.V.P., so an auxiliary view is one projected at right angles to an auxiliary plane. In other words, a view formed by taking sight lines parallel to the arrow ' A '.

Note also that the ground line of the auxiliary plane is marked $X_1 Y_1$. When an auxiliary view is drawn it is often referred to as a ' change in the ground line ', and the new ground line is usually called $X_1 Y_1$.

Fig. 171 shows the method of obtaining the auxiliary view of Fig. 170. The elevation is shown with the original ground line XY, and either the angle between XY and the new ground line (marked θ) is given, or the direction of the arrow marked ' A '. The former means that a view is required to be drawn from the new ground line, while the latter means that a view is required as seen in the direction of the arrow, that is, at right angles to it.

Now study the method carefully.

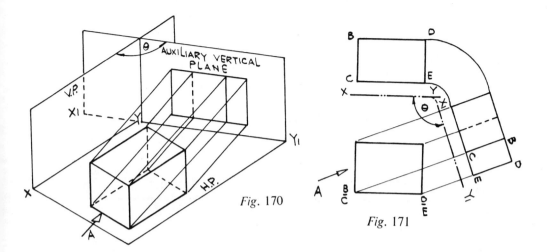

Fig. 170

Fig. 171

For an auxiliary elevation projectors are always drawn from the PLAN. Many students find this confusing at first, but when it is remembered that projections for a front elevation are taken from the plan in orthographic views, the fact can be more easily accepted.

Projectors, then, are taken from the plan. These projectors must always be at right angles to the new ground line $X_1 Y_1$. The vertical height of each corner of the block from XY is then transferred to $X_1 Y_1$ along the corresponding projector of that corner. Only four corners have been lettered in the figure. The height of B from XY in the elevation is marked from $X_1 Y_1$ on the corresponding projector, and the height of C from XY on the same projector. The heights of D and E, which will be the same in this case as B and C respectively, are then marked on their projector. The heights of the other corners are also taken, and the points are joined to complete the view, including the hidden edge.

Here are the rules for drawing an *auxiliary elevation*.
1. Draw the ground line as given, or, when not given, at right angles to the arrow shown. Call it $X_1 Y_1$.
2. Draw projectors at right angles to $X_1 Y_1$ from all necessary points and corners on the plan.
3. Transfer the heights of these points from XY on to their corresponding projectors, working from $X_1 Y_1$. (This means that the height of any point above $X_1 Y_1$ in the auxiliary view must be the same as the height of the same point above XY in the elevation.)

In brief: *project* from the *plan*; take *heights* from the *elevation*.

Copy Fig. 171. Make the block twice the size shown and θ 120°.

To Draw an Auxiliary Elevation of a Hexagonal Prism (Fig. 172)

The prism is shown in the elevation resting parallel to the H.P. An auxiliary view is required in the direction of the arrow. Project a plan, then a new ground line perpendicular to the arrow (that is, 70° to the horizontal, or 110° to XY). Project at right angles to $X_1 Y_1$ from each corner of the prism. Make the height of A from $X_1 Y_1$ equal to the height of A from XY. Mark B and the other corners in the auxiliary elevation in the same way. Then complete the view.

Make a copy of Fig. 172 if each face of the prism is 20 mm wide and 25 mm high.

To Draw an Auxiliary Elevation of a Square Prism with One Edge on the H.P. and one Face at 60° to It (Fig. 173)

Draw the plan and elevation and the new ground line at 120° to XY. Project at right angles to $X_1 Y_1$ from all corners of the plan. Transfer the height of each corner from XY along the corresponding projector from $X_1 Y_1$, then complete the view as shown, with one edge on $X_1 Y_1$.

Carry out this exercise, assuming the prism to be 50 mm long and the end 25 mm square.

Figs. 174 and 175 are intended for more advanced students, but they can be carried out by the application of the rules for auxiliary elevations, and a careful notation—letters or figures—of the corresponding corners.

Notice particularly in Fig. 175 how the ends of the axis, the apex G and point H are located in the plan and auxiliary elevation.

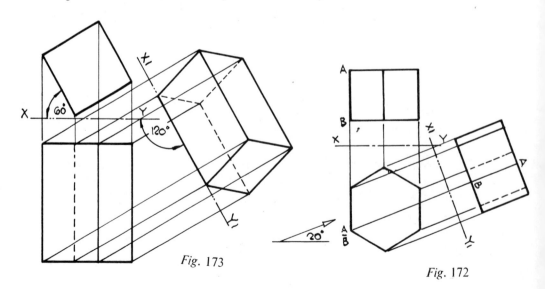

Fig. 173

Fig. 172

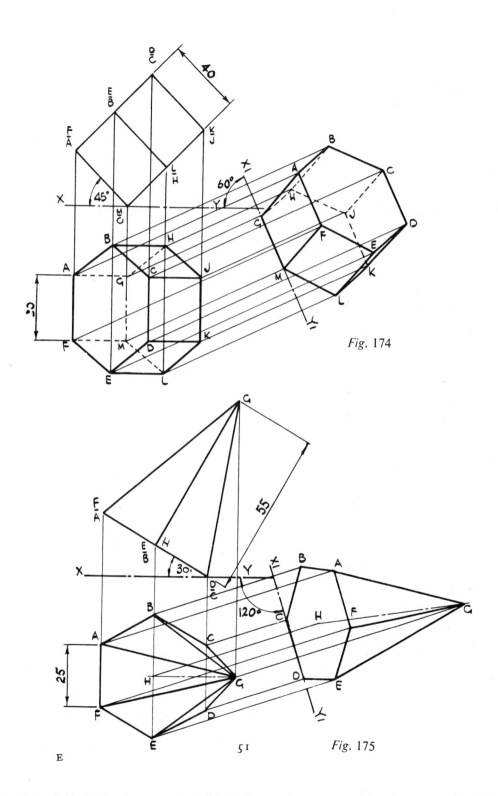

Fig. 174

Fig. 175

B. AUXILIARY PLANS

When the principle of auxiliary elevations is understood, there should be little difficulty in understanding the principle of auxiliary plans, and how to produce them.

In Fig. 176 an auxiliary horizontal* plane has been introduced at the angle θ to the H.P. The view of the rectangular block on this plane will be as seen in the direction of arrow ' B ', that is, at right angles to the auxiliary plane and outlined by the sight lines as shown.

The ground line for this new plane is also $X_1 Y_1$, while VT is the trace of the inclined plane on the V.P. and HT the trace on the H.P.

Fig. 177 shows how this view is obtained on the drawing paper. The plan and elevation are drawn, and the new ground line at the given angle. Projectors are then drawn at right angles to $X_1 Y_1$ from each corner of the ELEVATION. The distance from XY of each corner in the plan is then transferred to the corresponding projector from $X_1 Y_1$. In other words, the distance of A from XY is transferred to projector AA from $X_1 Y_1$, the distance B from XY to projector BB from $X_1 Y_1$. The other corners are marked in a similar manner and the respective points joined.

These are the rules for drawing an *auxiliary plan*.
1. Draw the new ground line and mark it $X_1 Y_1$.
2. Draw projectors at right angles to $X_1 Y_1$ from all necessary points and corners on the elevation.
3. Transfer the distances of these points in the plan from XY on to their corresponding projectors, working from $X_1 Y_1$. (This means that the distance of any point from $X_1 Y_1$ in the auxiliary view must be the same as the distance of that point from XY in the plan.)

In brief: *project* from the *elevation*; take *distances* from the *plan*.

If this is still found confusing, remember that as the plan is projected from the elevation in straightforward orthographic drawing, the auxiliary plan is likewise projected from the elevation.

Copy Fig. 177. Make the block twice the size shown, and $\theta = 60°$.

To Draw an Auxiliary Plan of a Hexagonal Prism Resting Parallel to the H.P.
(Fig. 178)

Draw the two views and the new ground line $X_1 Y_1$. Project at right angles to $X_1 Y_1$ from all corners of the elevation. Transfer the distance

---

*A term used here for clarity. Correct term: *auxiliary inclined plane*.

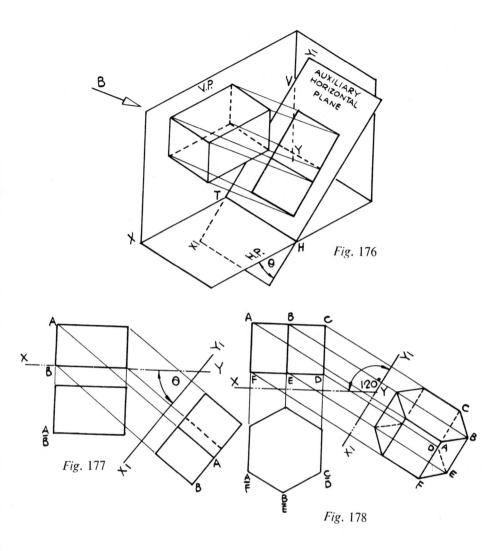

Fig. 176

Fig. 177

Fig. 178

of each corner from XY in the plan to the corresponding projector, working from X1 Y1 in the auxiliary elevation. Join the points correctly. showing hidden edges by means of broken lines.

Carry out the exercise of Fig. 178, assuming the prism to be 25 mm high with each side of the base 20 mm long. Some of the corners have been lettered to give some help. Compare your result with Fig. 172.

To Draw an Auxiliary Plan of a Square Prism Resting with One Edge on the H.P. and with One Face at 60° to that Plane (Fig. 179)

Draw the plan and elevation and introduce the new ground line. Project at right angles to X1 Y1 from each corner of the elevation. Take the distance of each corner from XY in the plan and transfer it to the corresponding projector in the auxiliary view, working from X1 Y1. Join up the appropriate points, indicating the hidden edge by a broken line.

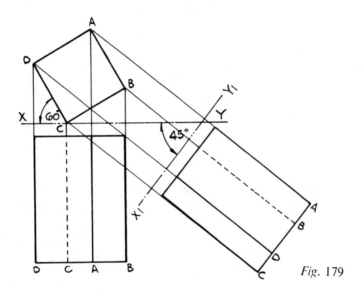

Fig. 179

Figs. 180 and 181 are harder exercises for keener students, who should be able to work them out by applying the rules for auxiliary plans. Fig. 180 is of a cylinder resting on the H.P. and can be drawn full size. Fig. 181 is of an octagonal pyramid and can be drawn to the given dimensions.

C. AUXILIARY SECTIONAL VIEWS

These views are obtained in a similar way to the other auxiliary views, except that the method is combined with that for obtaining sectional views. Two examples should be sufficient to make the method clear.

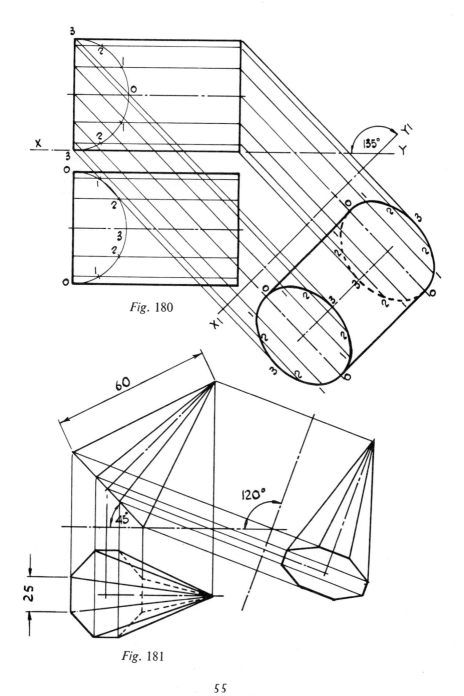

Fig. 180

Fig. 181

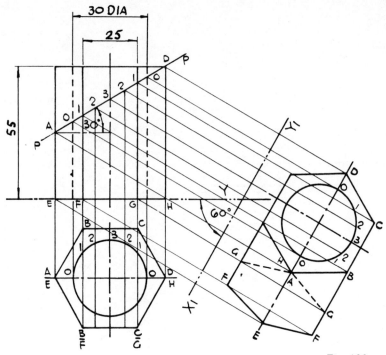

Fig. 182

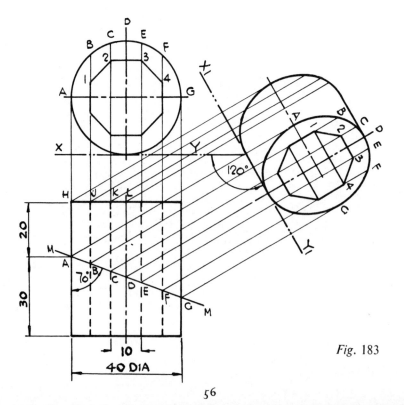

Fig. 183

To Draw an Auxiliary Plan of a Hexagonal Prism Pierced by a Cylindrical Hole and Cut by a Plane PP (Fig. 182)

Draw the two views in full, mark the cutting plane PP and the new ground line $X_1 Y_1$. Introduce traces in the plan and project these to PP. Then project at right angles to $X_1 Y_1$ from each of these points, A, 1, 2, 3 and B, and also from E, F, G and H on the base of the prism.

The distance of the points from XY is then transferred along the corresponding projectors from $X_1 Y_1$ as before, care being taken to ensure that the correct height is marked on the appropriate projectors. Join up the lettered points for the outline of the prism, and the numbered points for the section of the hole. The hidden edges of the cylinder have been omitted for clarity.

Draw Fig. 182 to the given dimensions. Show, wherever possible, all hidden edges.

To Draw an Auxiliary Elevation of a Cylinder Pierced by an Octagonal Hole and Cut by a Plane MM (Fig. 183)

Draw the two views, the cutting plane and the new ground line. Introduce traces in the elevation and project these to MM. Draw projectors at right angles to $X_1 Y_1$ from all marked points on MM, and also from H, J, K, L for the outline of the rear of the cylinder. Transfer the height of each point above XY to the respective projectors, working from $X_1 Y_1$. (Only the upper half has been marked for easier reference.) Join the lettered points for the outline of the cylinder and the figured points for the section of the hole.

Copy Fig. 183 full size, showing all hidden edges, if possible.

D. AUXILIARY VIEWS OF A STRAIGHT LINE

If a straight line is parallel to any of the three reference planes, its true length is shown by the view on that particular plane. If the line is parallel to the V.P. its true length is shown by the elevation; if it is parallel to the H.P. its true length is shown by the plan; and if the line is parallel to the S.V.P., by the end elevation.

Furthermore, if it is parallel to the V.P. or S.V.P., its inclination to the H.P. is correctly shown in the elevation or end elevation respectively, and if the line is parallel to the H.P. its true inclination to the V.P. is shown by the plan.

When a line is not parallel to any of these planes, as in Fig. 184, none of the views will show its true length or its true inclination to the H.P. and V.P. These things can, however, be worked out in several ways. Two of these methods will be dealt with here: the *orthographic method* and the *conical method*.

(i) The Orthographic Method

This method will be discussed first because, though it may not appear so easy, it follows on from the work in the previous sections, an auxiliary plane being introduced parallel to either the elevation or the plan, and the true length worked out on this plane.

Fig. 185(a) demonstrates this principle of introducing an auxiliary horizontal—it will now be called 'inclined'—plane parallel to the elevation. The distance of one end of the line (a_1) from the ground line XY is transferred to this plane, as is the distance of the other end of the line (b_1) from XY. When these two new points A and B are joined, an auxiliary view of the line is obtained, showing its true length.

Fig. 185(b) shows the method of applying this principle to the views of the line on the drawing paper.

Introduce a new ground line parallel to ab in the elevation. Draw projectors from a and b at right angles to $X_1 Y_1$. Transfer the distance of a_1 from XY (that is, $c_1 a_1$) to the projector aA, marking the distance from $X_1 Y_1$. This will give point A. Then transfer the distance of b_1 from XY (that is, $d_1 b_1$) to the projector bB, again marking the distance from $X_1 Y_1$. This will give point B. By joining AB the true length of the line is found.

Fig. 186(a) demonstrates a similar principle, but in this case an auxiliary vertical plane is introduced, being parallel to the plan of the line. The distance of a from XY is transferred to this plane from $X_1 Y_1$, and the distance of b from XY is transferred likewise. AB is then the true length of the line.

Fig. 186(b) shows you how to work this out when given the two views of the line.

Draw a new ground line parallel to the plan. Extend projectors from a_1 and b_1 at right angles to $X_1 Y_1$. Transfer the distance of the end a of the line from XY (that is, ac_1) to its projector, working from $X_1 Y_1$, then transfer the distance of the other end b from XY to its projector, working from $X_1 Y_1$. Join A and B for the true length.

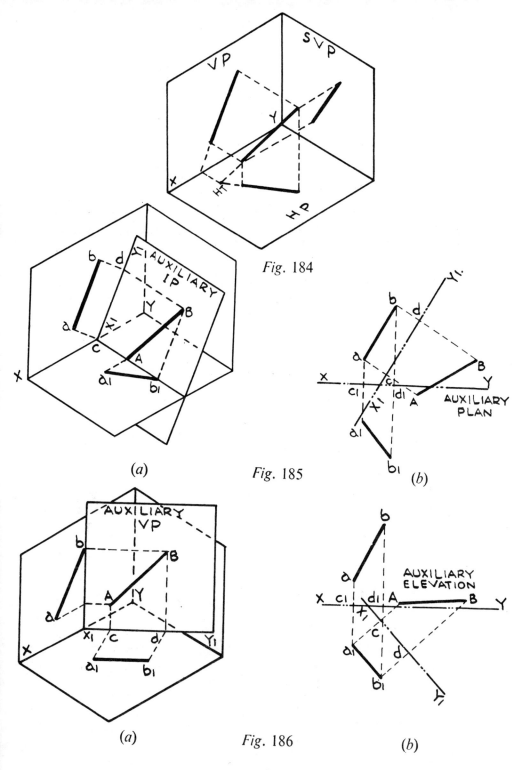

Fig. 184

(a) *Fig. 185* (b)

(a) *Fig. 186* (b)

To Draw the True Length of a Line and Its Inclination to H.P. and V.P. (Fig. 187)

Let the elevation and plan of the line be as in the figure. Obtain the true length of the line from both views.

You must now understand an important point. Just as an ordinary plan gives us the inclination of a line to the V.P., so an auxiliary plan will give us the true inclination to the V.P. You know that you project from an elevation for an auxiliary plan, so it is this view which will give the inclination to the V.P.

The method can be followed easily from the figure. Produce the true length line to the new ground line $X_1 Y_1$, and the angle between these lines is the true inclination to the V.P., namely, 25°.

Similarly, just as an elevation of a line gives its inclination to the H.P., an auxiliary elevation will give the true inclination of the line to the H.P. So produce the auxiliary elevation to the new ground line $X_2 Y_2$, and the angle between these lines will be the true inclination to the H.P., namely, 21°.

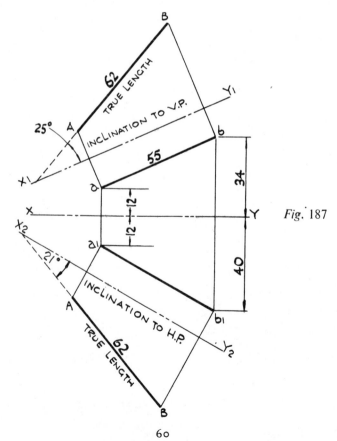

Fig. 187

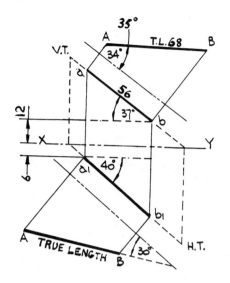

Fig. 188

Draw the elevation and plan of the line in Fig. 187 full size. Find its true length and inclination to both planes and check the result with that shown.

To Find the True Length of a Line, its True Inclination to Both Planes, and also its Horizontal and Vertical Traces (Fig. 188)

Work out the true length and inclinations of the line as above.

The trace of a line is a point. In other words, it is what we see if we sight exactly along the line when it is at eye level. The *horizontal trace* of a line is the point where the line (or more often, the line produced) meets the H.P. Refer to Fig. 184 and you will see the horizontal trace marked on that plane. The *vertical trace* of a line is the point where the line—or the line produced—meets the V.P. This is not shown on Fig. 184, but you could mark its position with your straight edge and you would find that it lies below the XY line and beneath the H.P.

To find the horizontal trace of Fig. 188, produce ab to the ground line XY, then drop a perpendicular to meet a_1b_1 produced. To find the vertical trace, produce a_1b_1 to the ground line XY, then erect a perpendicular to meet ab produced. The intersection of these lines will give the required trace in each case.

Work out the problem of Fig. 188 and compare the results with those shown.

Fig. 189 shows two rather interesting points. Firstly, it will be noticed that no new ground line has been introduced to find the true length. This is a method of saving time in working out these problems. The new ground line is really ab. Perpendicular projectors are drawn to ab and the distance of a_1 from c (not from XY) is marked along the projector from a to give point A. Since b_1 already lies on this imaginary ground line cb_1, it will coincide with b in the elevation. The auxiliary elevation is obtained in the same way, working from the line bd. This method is quite sound, providing it is not confused with the complete method already explained.

The second point regarding Fig. 189 is in connection with the horizontal trace. Notice that it does not appear to lie in the H.P. It is found by producing ab to XY as before. Since, if a perpendicular is dropped from this point into the H.P., it will be moving away from the direction of a_1b_1, it is necessary to double back, as it were, ab produced by erecting a perpendicular into the apparent V.P. until it meets a_1b_1 produced. The intersection of these two lines will then give the horizontal trace.

This does not mean, however, that the horizontal trace is in the V.P., for this cannot be. a_1b_1 is actually penetrating the V.P. 12 mm above XY, and comes to rest on the H.P. the other side. A similar case occurs in Fig. 184, where the vertical trace comes to rest on the V.P. below the H.P.

Since all the orthographic drawing so far shown in this work is in first angle projection, it is said to be drawn in the *first quadrant*, which is the quadrant shown in Figs. 184, 185(a) and 186(a), as well as in other figures.

The horizontal trace of the line in Fig. 189 is therefore said to be in the *second quadrant*, while the vertical trace of Fig. 184 is said to be in the *fourth quadrant*, since the quadrants are numbered in an anti-clockwise direction (see Part III, Chapter 5).

If this is found difficult to understand, the student should not be unduly worried. It is more essential for the immediate purpose, that he can find the horizontal and vertical traces of any given line.

(ii) The Conical Method

Once this method is understood, it will probably be preferred to the previous one.

Suppose the elevation of a line is ab, the plan a_1b_1, and the true length AB, as in Fig. 190.

If the figure is studied carefully it will be seen that the line AB, the line

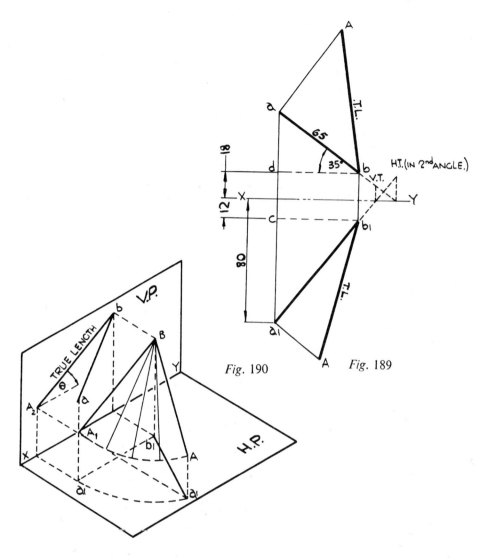

Fig. 190   Fig. 189

a_1b_1 and their projectors to each other, form a quadrilateral $ABBb_1a_1$. Imagine this quadrilateral to be now slowly turned round, with b_1B as the hinge, until it is parallel to the VP. The impression of this quadrilateral can now be cast on to the V.P., and the line A_2b so shown will be the true length of ab, while the angle θ will be its true inclination to the *horizontal* plane.

What has really been done, is to make the line AB the slant side of part of a cone, with B as its apex. Hence the name conical method.

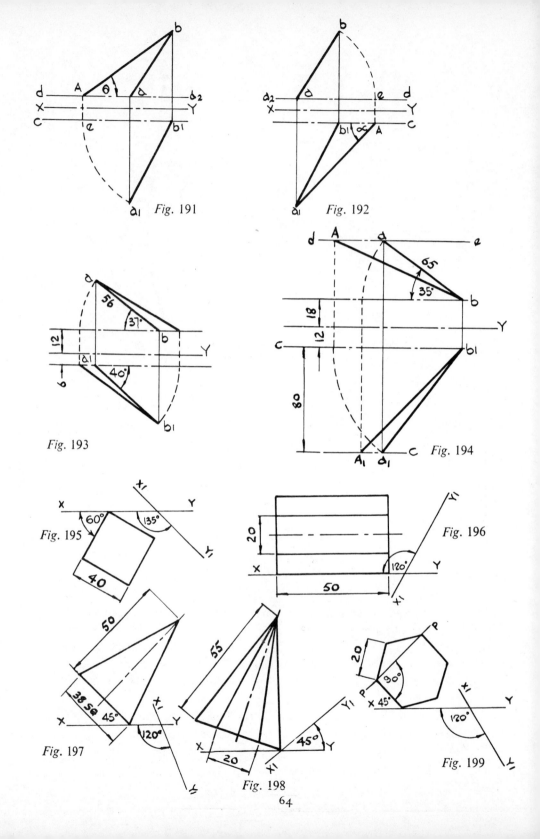

Fig. 191

Fig. 192

Fig. 193

Fig. 194

Fig. 195

Fig. 196

Fig. 197

Fig. 198

Fig. 199

Fig. 191 shows the method of carrying this process out on paper. Given the two views, draw a2d parallel to XY (through a) and b1c parallel to XY. With centre b1 and radius a1b1 describe an arc to cut b1c at e. Project point e perpendicularly to cut ad at point A. Join Ab for the true length of the line. θ gives the true inclination of the line to the H.P., for Ab is an elevation on the V.P. and NOT an auxiliary view on an auxiliary plane.

Fig. 192 shows the method of finding the true length of the same line by reference to the H.P., and therefore, also, the true inclination of the line to the V.P.

Draw a2d and b1c parallel to XY as before. With centre a and radius ab describe an arc to cut ad at e. Project e vertically to cut b1c at point A. Join A to a1 for true length. The Greek letter α (alpha) indicates the angle Aa1 makes with the vertical plane.

The traces of the line, of course, are found in the same way as before.

Figs. 193 and 194 are the same problems as Figs. 188 and 189, but they are worked out by the conical method.

Fig. 194 should give no difficulty if the rule is remembered: The centre of the arc is nearer XY; the perpendicular projector is to the opposite end (in the other view).

Carry out the exercises of Figs. 193–4. Compare the results with your answers to Figs. 188–9.

In some problems the true length of a line is given and only limited information about the elevation of the line and/or the plan. For instance, if in Fig. 194 the true length Ab was given together with the exact position of the plan, work backwards to determine the elevation of the line. Draw line de through A and parallel to XY, and project a1 to this line to find the position of a in the elevation. Other variations can be worked out in a similar way.

EXERCISE 4

1. Construct an auxiliary elevation of the cube shown in Fig. 195 on the plane X_1Y_1.
2. Fig. 196 shows a hexagonal prism resting on the H.P. Construct an auxiliary plan on X_1Y_1.
3. Draw an auxiliary view on the new ground line of Fig. 197.
4. A hexagonal pyramid is shown in Fig. 198. Draw an auxiliary plan of the pyramid on the ground line shown if its base is a 20° to XY.
5. A hexagonal prism 50 mm long rests with one edge on the H.P. as shown in Fig. 199. Draw an auxiliary sectional elevation of the prism on X_1Y_1, when it is cut by a plane PP.

6. A hollow cylinder is cut by a plane PP as shown in Fig. 200. Draw an auxiliary sectional plan of the cylinder on X_1Y_1.
7. A square prism rests in a central position on an octagonal prism 50 mm long as in Fig. 201. Draw an auxiliary elevation of the two prisms in this position on the ground line X_1Y_1.
8. Two views of a straight line are shown in Fig. 202. Find by construction the true length of this line, its inclinations to the V.P. and H.P., and the position of its horizonatal trace.
9. Find the true length of the line shown by the two views in Fig. 203. Find also its inclination to the H.P. and vertical trace.
10. Find the true length of the one shown in Fig. 204, its inclinations to both planes, and also its horizontal and vertical traces.

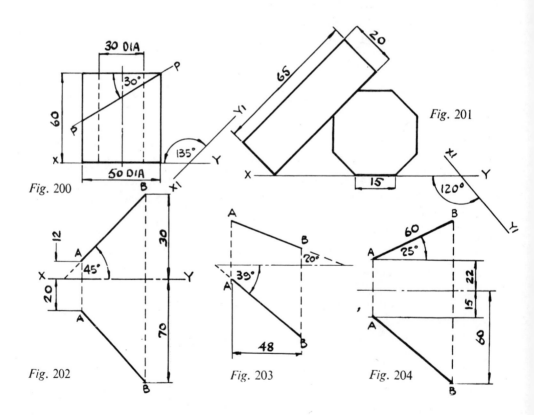

Fig. 200

Fig. 201

Fig. 202

Fig. 203

Fig. 204

CHAPTER SEVEN

Intersection of Solids

Many single objects are made up of two geometrical solids formed together so that their respective shapes merge into each other. They are then said to *interpenetrate*. When this occurs, as it does frequently in building and engineering, the line or curve where the two surfaces meet has to be worked out before the object can be made. This line is known as the *line of intersection*, and is of great importance.

Beginning with simple solids, we will see how this line or curve can be shown on the drawing paper.

Fig. 205 shows two views of an equilateral triangular prism joined to a larger horizontal prism. The outline of the two views is drawn with the aid of the auxiliary plan ABC.

The edges are then lettered in each view to correspond. This lettering—or figuring, as the case may be—must be done most carefully as in previous work. The point where edge B meets the large prism in the end elevation is projected horizontally to meet edge B in the elevation. The point where edge A meets the prism is also projected to edge A in the elevation. These two points are then joined to give the elevation of the line of intersection.

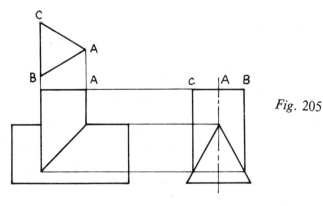

Fig. 205

Simple as this may seem, it is the basis of all work on the interpenetration of solids.

These are the rules for guidance:

1. Draw two views of the object in projection, making sure that all edges are in their relative positions.
2. For circular objects, having no edges, introduce traces as in Fig. 146 etc.

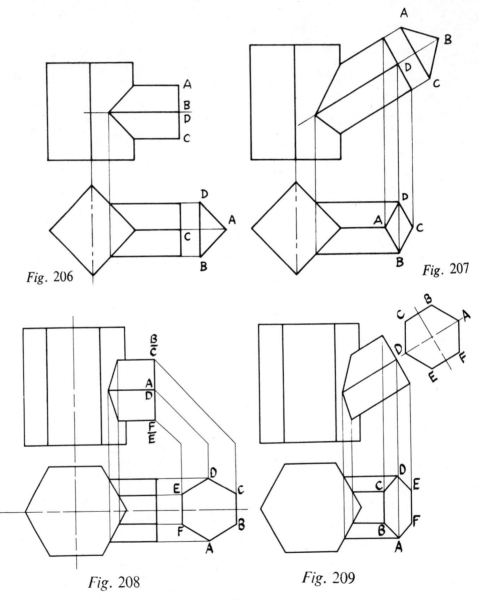

Fig. 206

Fig. 207

Fig. 208

Fig. 209

3. Letter or number the edges or traces carefully to correspond in each view.
4. Project the points of intersection in one view to the corresponding edges or traces in the other.
5. Join the points so marked with a line or smooth curve as required.

It should be easy to copy Figs. 206–9. Letter each view carefully, and draw any necessary auxiliary views.

Fig. 210 shows an octagonal prism resting vertically on a triangular prism. Draw the end elevation first, using the auxiliary half plan to obtain the edges, then project for the elevation. Letter the edges carefully, and project the points of intersection to the corresponding edges as before. Join the points in the elevation as shown.

Fig. 211 shows the same prisms, but the octagonal prism is now set obliquely on the other. Copy the drawing, making the oblique prism 30° to the horizontal.

Fig. 212 needs further explanation. The object consists of a cylinder resting vertically on a triangular prism. A semicircle is described on the

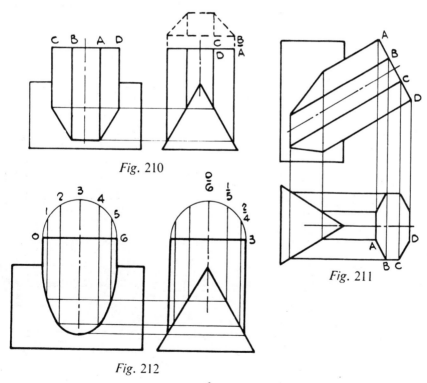

Fig. 210

Fig. 211

Fig. 212

cylinder in each view, then both are divided into six equal parts with the aid of the 60° set-square. Traces are then drawn through the points on the semicircles and are carefully numbered.

Number the elevation first. Now suppose the semicircle on this view is hinged at O and 6, then pulled towards you until it is in a horizontal position. A view from the left of the elevation will show the figures as numbered in the end elevation. Mark these as shown.

The rest of the construction follows as before. Always project the limits of the curve—that is, traces O, 6 and 3—first. These act as a useful guide.

Fig. 213 is carried out in the same way. Once the traces are correctly numbered, the rest of the work is straightforward. Once again, project traces O, 6 and 3 first.

Fig. 214 is really a similar construction. Again, imagine the semi-circle in the elevation to be hinged towards you, in order to number the traces in the plan to correspond. Let the axis of the oblique cylinder be at 30° to the horizontal, and as a further exercise, try to produce an end elevation.

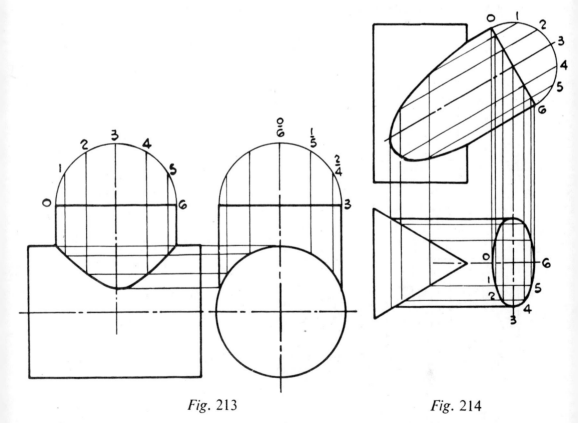

Fig. 213 *Fig. 214*

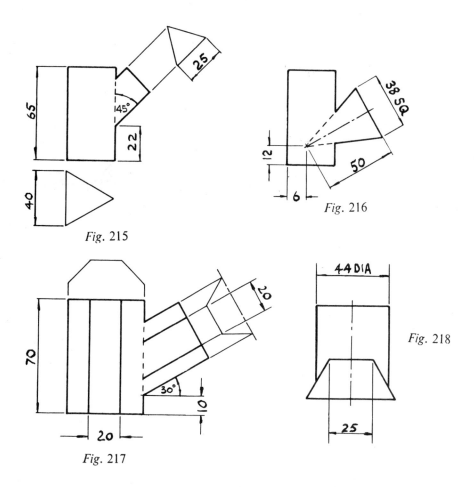

Fig. 215

Fig. 216

Fig. 217

Fig. 218

EXERCISE 5

1. Draw the line of intersection of the two triangular prisms shown in Fig. 215, each being equilateral in cross section.
2. The vertical prism of Fig. 216 is of the same dimensions as that of Fig. 215, but is joined to a square pyramid. Draw an elevation of the intersection of the two solids, if the axis of the pyramid is at 30° to the horizontal.
3. A hexagonal prism is joined to an octagonal prism as shown in Fig. 217. Draw an elevation of the intersection of the two prisms.
4. A cylinder rests on a half hexagonal prism as shown in Fig. 218. Draw the curve of intersection, assuming the prism to be 65 mm long.

5. An end elevation of two cylinders joined together is shown in Fig. 219. Draw the curve of intersection, assuming the missing lengths.
6. Two cylinders are joined obliquely as in Fig. 220. Draw the curve of intersection of the cylinders.

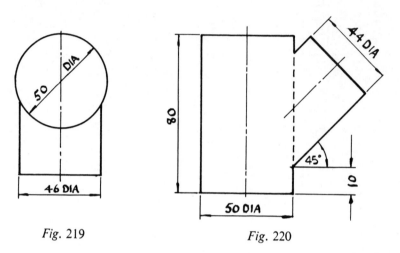

Fig. 219   Fig. 220

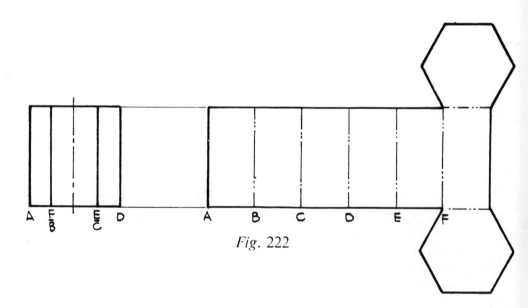

Fig. 222

CHAPTER EIGHT

Development of Surfaces

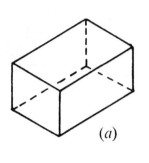
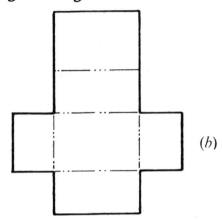

Fig. 221

The development of a geometrical body or other object is the shape in the flat of the surface of the object. Match boxes, cigarette packets, cartons and many articles made from sheet metal begin as flat surfaces cut to the required shape and size, usually with the aid of a template.

This template is a full-size pattern of the required object worked out from a drawing. Sheet metalworkers and steelworkers often have to develop the surface in the flat of shaped parts on special boards fixed to the floor of the workshop. In such cases laps or some other allowances are often made for purposes of fixing and in geometry these allowances are called flaps. They are usually omitted, however, in problems on surface development, and there is no need to include them unless specifically requested.

If it is required to develop a shape that will form a cube, it should be easy to realize that this will merely be six squares joined together, equal in size to one side of the cube.

Fig. 221(a) shows the rectangular model used frequently before. Fig. 221(b) shows the development of the surface of this model, the chain-dotted lines indicating the bend lines of the adjacent parts.

A hexagonal prism and its development is shown in Fig. 222. It can be seen that the shape of the surface is merely a rectangle of the same height as the prism and of length equal to six times the length of one side, with a top and bottom adjoining.

To Develop the Surface of a Prism Cut by a Plane (Fig. 223)

The same method applies to any regular prism.

Draw an auxiliary plan—or half plan, as in the case of the figure—and project from it an elevation. Letter the plan and mark the edges in the elevation to correspond. For the development project a base line GG from the elevation, making it six times AB in length, then divide it into six equal parts. Erect perpendiculars to GG and letter these lines successively as in the plan.

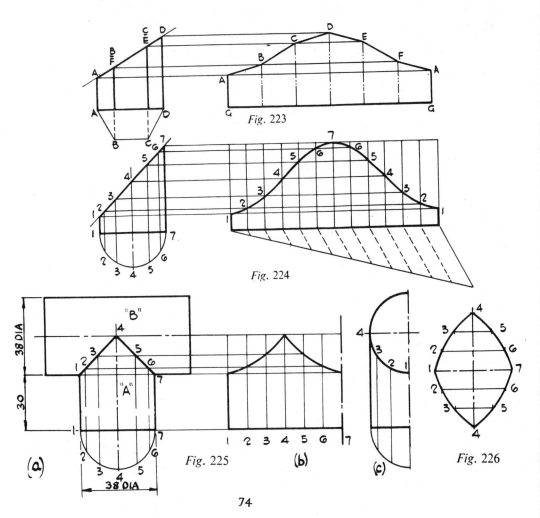

Fig. 223

Fig. 224

Fig. 225

Fig. 226

It is a rule in engineering that all joints should be made on the shortest line when possible, mostly to save materials, and it is a good rule to adopt in geometry.

Edge A is the shortest in Fig. 223, so it has been chosen for the joining edge. The height of each edge is now projected across to its corresponding edge in the development. Once again, it must be stressed that care is necessary in lettering. These heights are then joined by straight lines as shown. The development of the base and top have been omitted because their shapes are usually easy to draw, and generally are not required in this work.

To Develop the Surface of a Cylinder Cut by a Plane (Fig. 224) *or Planes* (Fig. 225)

As the development of a plain cylinder will be a rectangle equal to the height of the cylinder and of length equal to its circumference, so the length of the development of any cylinder, no matter how it is cut—provided both ends are not involved—will be equal to the circumference of the cylinder.

The elevation is drawn, an auxiliary half plan is described on the base, then this is divided into a number of equal parts, preferably six. Traces are then projected to the cutting plane from these points and numbered as in the plan.

Project a base for the development equal to the circumference of the cylinder, using the formula $c = \pi d$. It is more accurate to take π as 3·14 for this purpose. The base line is divided into 12 equal parts for the whole development, either in the manner shown or by continued bisection, then perpendiculars are erected from these points and the traces carefully numbered. The heights of the corresponding traces in the elevation are projected to the development, and the points so obtained joined with a smooth curve.

Construct the development of a cylinder as in the figure, making the diameter half as large again as that shown.

Fig. 225(a) shows two cylinders of equal diameter intersecting at right angles. A *half* development is shown at (b) and its construction should be easy to follow. Note that the traces are numbered below the base line. Previously they were numbered near their intersections to help in following the construction, but the method of Fig. 225 is the more usual.

To Develop the True Shape of an Intersection in a Cylinder (Fig. 226)

By comparing Fig. 225(a) with its half end elevation at (c), it should be obvious that the length of the centre line of the hole will, in this case, be equal to half the circumference of the horizontal cylinder, that is, twice the length of the arc 4–3–2–1 in the end elevation of Fig. 225. This can be obtained by dividing the vertical cylinder into a number of equal parts—six in Fig. 226 for the complete end elevation, but the more that are used the more accurate the result will be—and erecting traces from these points to the horizontal cylinder. The distance from 1 to 2 can now be stepped out six times along the centre line 4–4, to give the length of the complete arc.

Horizontal traces are drawn through these points and numbered to correspond with the elevation. Working from the centre-lines in each case, the lengths of the horizontal traces in the elevation are transferred to their respective traces on the true shape. The points so formed are then joined with a smooth curve.

Make a complete development of cylinder 'A' in Fig. 225, and a true shape of the hole in cylinder 'B'.

To Develop the Surface of a Square Pyramid (Fig. 227)

A close study of the elevation of this figure will show that the edges in the elevation are not true lengths. An important rule is:

Developments can only be worked out from lines of true length.

The true length of the edge is found by the conical method (page 62). With centre F and radius FB describe an arc to the horizontal centreline for point E, project E to the base of the elevation, then join FE.

What is really done is to turn the pyramid round until the edge is parallel to the V.P., as shown in the auxiliary views, where the true length of FB can now be measured.

To work out the development, strike out an arc of radius FE (or FB in the auxiliary elevation). As the length of each base edge in the plan is a true length, the distance AB is marked off four times round the arc. The points are joined in order to form the base edges, then linked to F for the bend lines.

The square forming the base has been purposely omitted, as it will be in all subsequent examples.

Make a copy of Fig. 227 full size.

To Develop the Surface of the Frustum of Any Pyramid (Fig. 228)

The method is exactly as above, but the arc for the development has been drawn with the apex of the pyramid as its centre. This saves time when space is available on the paper.

To complete the development, project the upper edge of the frustum (line 1–2) to the true length line and draw a second arc of radius F1. Join the points where this arc intersects the bend lines.

Construct the development of the truncated pyramid in Fig. 228, without reference to the method.

Fig. 229 shows a pyramid cut obliquely. Mark all corners systematically in the plan—the method shown being a sound one—and the upper corners in the elevation to correspond. To obtain the true lengths, project 2 and 3 in the elevation to the true length line EA, and also points 1 and 4. Describe arcs of these radii to the correct bend lines, then join up the points so indicated with straight lines.

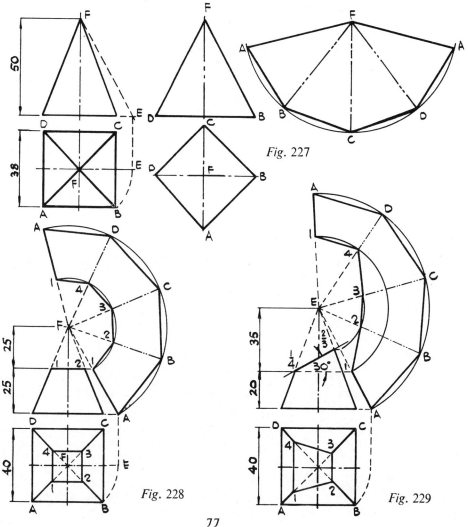

Fig. 227

Fig. 228

Fig. 229

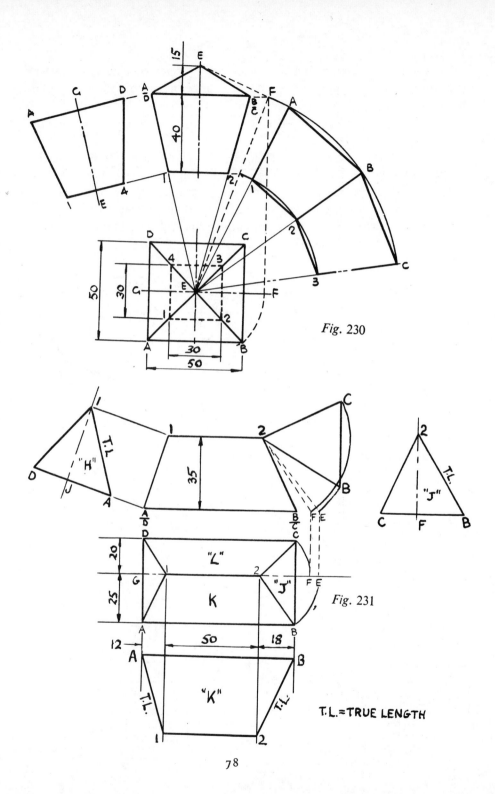

Fig. 230

Fig. 231

T.L.=TRUE LENGTH

Again, all heights in the elevation must be transferred to a line of true length.

Work out the developments of Fig. 229 as shown, using the same notation for all corners.

Fig. 230 is a problem which may be met in examinations. It is required to develop the surface for a lantern, as shown in the elevation and plan.

If space permits, it is a good method to draw the plan in the position shown, its centre E being also the apex of the lower parts of the model. It can be seen that the problem is really of two pyramids, and becomes simple to solve when the true lengths have been worked out. The development of the upper part has been omitted because it is quite straightforward and only a half development of the lower part is shown.

If the taper of the sides is so slight that the apex cannot be found, the development can be worked out as a true shape (Chapter 5), which of course it is. A centreline EG is drawn parallel to 1A of the elevation and the vertical distances on the plan are marked along perpendicular projectors as shown on the left of the elevation. If additional sides are required, they can be constructed from the first one by marking off AD and 1–4 from A and 1, then using a diagonal to cut the arcs.

To Develop the Sides of a Hipped Roof

Fig. 231 is an additional problem. It shows two diagrams of a hipped roof with its ridge out of centre, and it is required to draw a pattern of the sides ' G ' and ' H '. This can be carried out by the second method of Fig. 230, as shown on the left of the elevation.

It can also be worked out as shown on the right of the elevation of Fig. 231. In this case it must be realized that the line 2B in the elevation is a true length of the centre line 2F in the plan, a fact already dealt with in the work on true shapes, and the distances of C and B are then transferred from the plan.

Finally, the problem can be solved by the use of true lengths. The true lengths of 2B and 2C (2E and 2F respectively, in the elevation) are found, then arcs of these radii are described. A suitable point B is chosen on the larger arc and the distance BC in the plan is marked from B on to the other arc. Points 2, B and C are then joined for the pattern.

If it is desired to develop side K, this can be solved by projecting lengths AB and 1–2 from the plan and using the true length of 2B to locate point 2 of K, as shown, and the true length of 1A to locate point 1 of K. Side L can be determined in a similar manner.

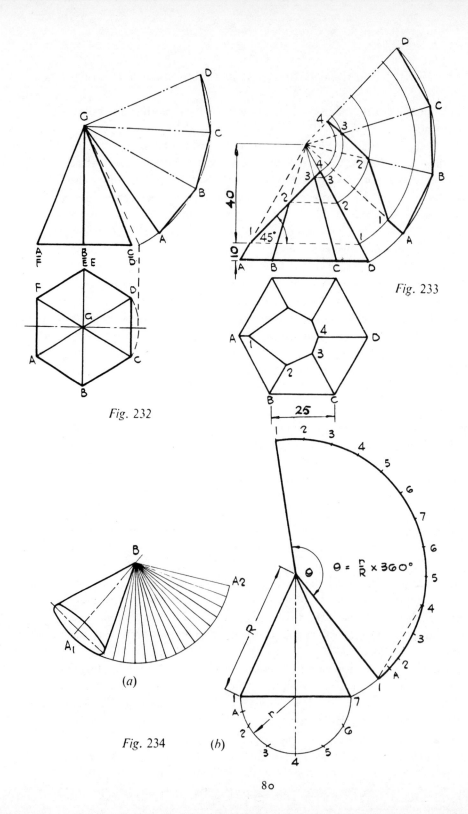

Fig. 232

Fig. 233

Fig. 234

To Develop a Hexagonal Pyramid

Fig. 232 shows a half development of a hexagonal pyramid, and should be easy to follow.

Fig. 233 should also be self-explanatory. Note that in this case 4D is parallel to the V.P., and is therefore a true length in the elevation. Points 1, 2 and 3 are projected to it, then the respective arcs are described as before.

To Develop the Curved Surface of a Cone (Fig. 234)

If a cone is placed on its side with point A_1 in contact with the H.P. and then rolled round carefully until the point is in contact with the H.P. again at A_2, the shape so defined by the sector BA_1A_2 will be a development of the curved surface of the cone.

This is the principle of developments on cones, and the only difficulty in the early stages is that of working out the correct length of the arc, which is of radius equal to the slant height.

Fig. 234(*b*) shows an accurate method when this dimension R is given or can be calculated, for the formula shown can then be used for the angle θ. For example, if $r = 25$ mm and $R = 50$ mm, then the angle will be:

$$\frac{25}{50} \times \frac{360°}{1} = 180°$$

An approximate method is to draw a half plan on the base of the cone and divide it into a suitable number of parts; the greater the number the more accurate, again, the result will be. Twice the number of divisions are then stepped round the arc, because the semi-circle shows only half the circumference.

There is no need to step out each division separately. One division can be sub-divided for greater accuracy, as at A. 1A and 2A are then marked out in turn on the arc, two more divisions marked off equal to 1–2, then the chord from 1 to 4 marked round the required number of times. The ends of the arc are joined to its centre (the apex of the cone, in this case) to complete the shape. The numbering of the divisions round the arc is unnecessary in this example but it should be understood.

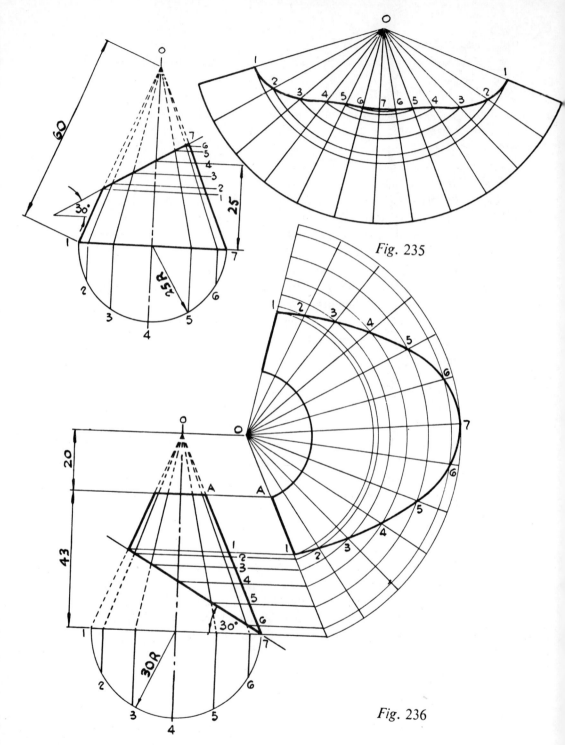

Fig. 235

Fig. 236

To Develop the Curved Surface of the Frustum of a Cone (Fig. 235)

Draw a semi-circle on the base of the cone, divide it into six equal parts and number them. Project each point to the base, then join it to the apex.

The traces are not true lengths, so the point of intersection of the cutting plane with the trace must be projected to the line of true length, O_7. From a suitable point (the apex O can be used), describe an arc of radius O_7 and complete the sector by drawing the required angle at O. Divide the arc into 12 equal parts, bisecting till one quarter is found, then trisecting this by trial and error. Join each point to O and number carefully.

The true length of each trace is now taken from O_7 in the elevation and marked from O on the respective trace on the development. Finally the points are joined with a neat curve.

Construct the development of Fig. 235 as shown.

To Develop the Curved Surface of a Cone Cut by Two Planes (Fig. 236)

Suppose a truncated cone has to be made to fit a sloping roof as shown in the figure. The method of developing the curved surface is clearly shown, and will be left without explanation, as an exercise.

(Note: As practical exercises any of these developments can be cut out and folded to shape to prove the construction.)

EXERCISE 6

1. Fig. 237 shows a diagram of a hexagonal pipe fitted to the sloping roof of a tank. Construct a development of the pipe.
2. An octagonal prism is cut by a plane as at Fig. 238. Draw a development of the lower portion.
3. Fig. 239 shows a cylinder cut by a plane. Draw a development of the upper portion.

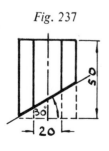

Fig. 237

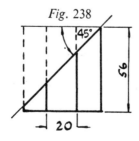

Fig. 238

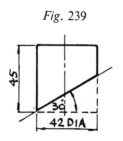

Fig. 239

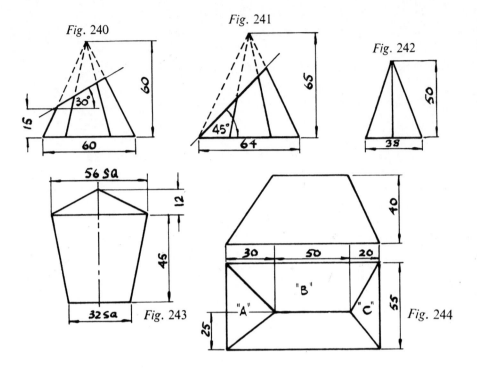

4. A sheet-iron ventilating duct 76 mm in diameter has a 70 mm branch entering at right angles. Give a development of the hole which must be cut in the 76 mm diameter pipe.
5. Construct the pattern required to make the frustum of the hexagonal pyramid shown in Fig. 240.
6. An octagonal pyramid is cut by a plane as in Fig. 241. Draw the development of the frustum.
7. Fig. 242 shows an elevation of an equilateral triangular pyramid. Construct its development.
8. An elevation of a lantern is shown in Fig. 243. Draw a complete pattern of both parts.
9. A diagram of a hipped roof with its ridge out of centre is shown by the two views in Fig. 244. Draw developments of parts 'A', 'B' and 'C'.

10. Fig. 245 is an elevation of two cylinders meeting obliquely. Make a full development of cylinder 'A'.
11. A right cone is cut by an oblique plane as at Fig. 246. Draw a development of the frustum.
12. Fig. 247 shows the diagram of a conical vent fixed to a sloping roof. An elliptical outlet is attached to the vent as shown. Draw a complete development of the vent.

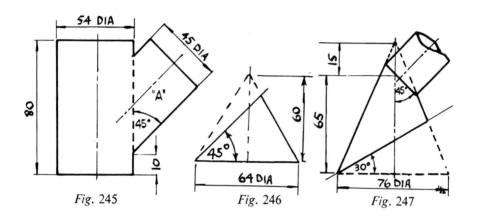

Fig. 245   Fig. 246   Fig. 247

Test Papers

These tests have been chosen from G.C.E. papers set by the Examining Boards stated to cover as wide a field as possible. The conversion of dimensions to approximate metric equivalents has also been undertaken with permission.

All methods of construction must be shown clearly for each question and they must be geometrical methods whenever possible. The student could be given a choice of two questions from each three, as with previous test papers, and about one hour would be reasonable for each test.

TEST 1

1. Fig. 248 indicates two cylinders which meet at 45°, their axes being in the same vertical plane. Find the elevation of the curve of intersection of the cylinders; also the development of the surface of the cylinder 'A'. The joint is to be along the line XX.
(Univ. of London)

2. Fig. 249 shows the elevation of a 44 mm cube with a 24 mm square hole passing centrally through it. Draw the elevation and plan of the solid. Project from the plan an auxiliary elevation on the ground line X_1Y_1. Show all hidden edges.
(Univ. of London)

3. The wedge-shaped solid shown in Fig. 250 is made by machining two flats on a 76 mm length of metal of diameter 64 mm. At right angles to the wedge a slot 25 mm wide by 25 mm deep is machined.

Make an isometric drawing of the wedge in the direction S. Do not use an isometric scale.
(Univ. of London)

TEST 2

1. Draw, full size, an oblique parallel projection of the bracket shown in Fig. 251. The plane of symmetery is to remain vertical and is to be parallel to the angle of inclination. The top of the slot should be completely visible.
(Oxford Local Exams.)

2. Fig. 252 shows a cylinder from which the part above the horizontal plane has been removed. Draw the plan of the solid. No hidden edges need be drawn.
(Univ. of London)

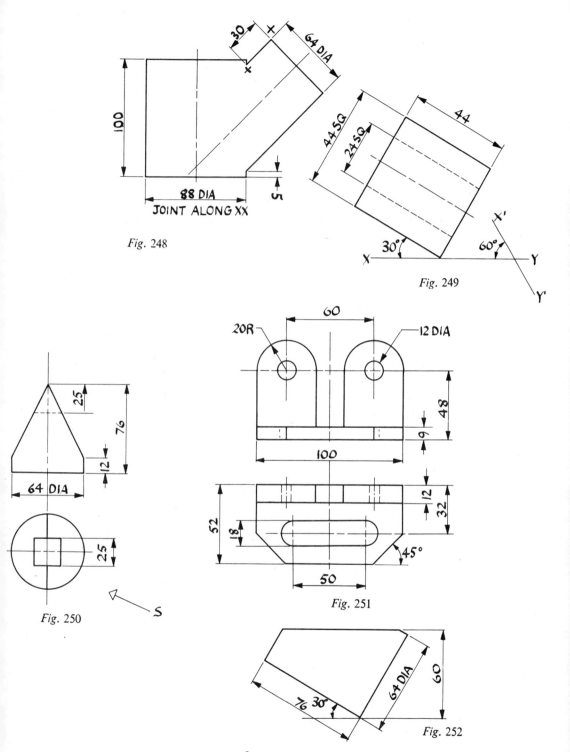

Fig. 248

Fig. 249

Fig. 250

Fig. 251

Fig. 252

3. Fig. 253 shows the plan of two straight lines which meet at B, also the elevation of them. The true lengths of the lines are the same. Draw the elevation of the line BC, given that C is above the H.P. and find the true angle between AB and BC.   (Univ. of London)

TEST 3

1. Fig. 254 shows a plan and an elevation of the top of a small wheel-barrow made of sheet steel. Draw:
    (i) the given elevation;
    (ii) the plan;
    (iii) an end elevation looking from the left;
    (iv) a plan looking in the direction of ' A ';
    (v) the true shapes of the back, front and one side.
   Ignore any allowances for joints. Scale 1 : 4.
   (Oxford Local Exams.)

2. Fig. 255 shows two views of a casting. Draw these views and add a plan, also an end elevation in the direction of the arrow-head ' S '. No hidden edges need be shown.   (Univ. of London)

3. The elevation of a piece of T-section bar is shown in Fig. 256. Draw the following views of the bar:
    (a) the given elevation E;
    (b) a plan projected from view (a);
    (c) an auxiliary elevation on a plane the horizontal trace of which is X_1Y_1.   (Univ. of London)

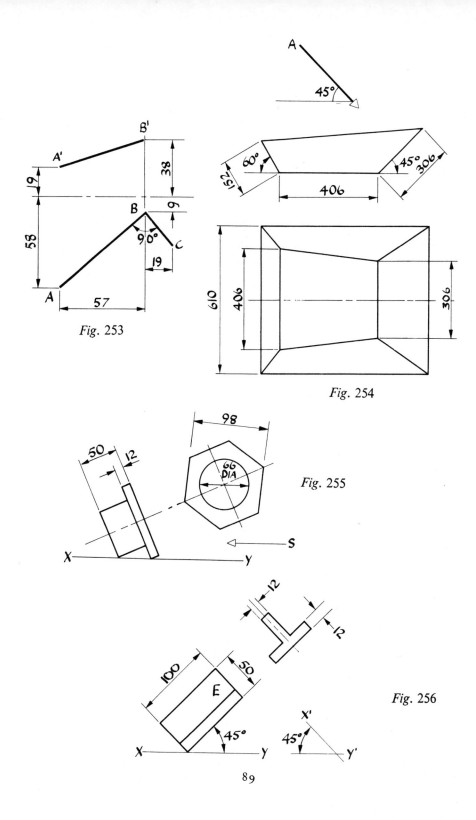

Fig. 253

Fig. 254

Fig. 255

Fig. 256

PART III

Technical Drawing

CHAPTER ONE

Methods of Fastening

Technical drawing and engineering drawing are a continuation of the orthographic drawing already discussed, except that the objects drawn are those applicable to industry, and particularly to engineering, and are more detailed than the geometrical models drawn so far.

Orthographic projection is more commonly used for a working drawing, that is, a drawing from which the object can be worked or made, and though third angle projection (see Chapter 5) is adopted by many firms, first angle projection is still popular and will be used first since it follows on naturally from Part II.

It is important to know something of the methods of fastening in common use before proceeding to the drawing of parts, for all engineering products, when complete, are made up of various items, or components, fitted (or assembled) together.

A common method of fastening parts together is by *screw fastenings*, which can take the form of bolts and nuts, screws or studs (see Figs. 257 and 258). These are available in various standard sizes and lengths, and have the advantage that the fastening can be readily detached for adjustment, repair, replacement etc. by removing the nuts or screws to release the parts.

A *bolt* has a cylindrical *shank* (normally parallel) and a *head*. The free end of the bolt is *threaded*, the rest of the shank up to the head being left plain, i.e. unthreaded. The length of the threaded portion is about twice the diameter of the bolt and on it is screwed the *nut*, which is *tapped*, or threaded

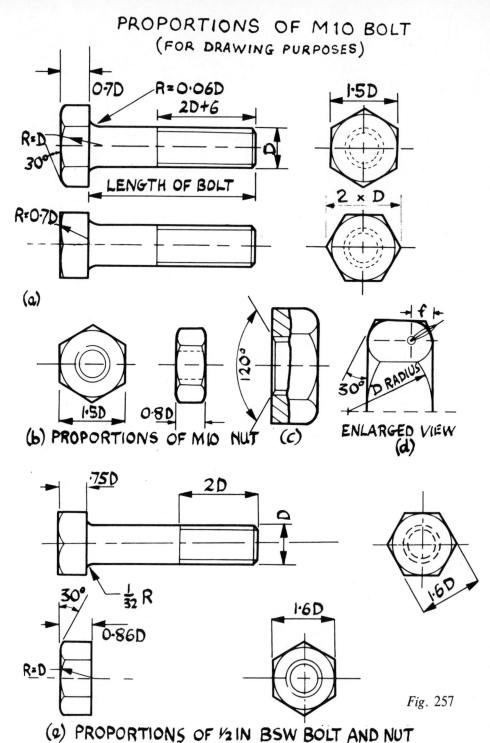

Fig. 257

internally, to suit the bolt. There are several standard types of bolt head, e.g. countersunk head, square head, snap or round head (see Fig. 259(a)), but the commonest type, particularly for precision (bright) bolts, is the hexagon head.

Fig. 257(a) shows the standard proportion for a metric precision bolt of 10 mm diameter as specified by the ISO (International Standards Organization). This bolt size is known as M10, M to denote metric thread and 10 its diameter. The proportions of smaller and larger bolts may differ slightly from these, but, for the purposes of engineering drawing, the differences are small and the proportions given may be generally used for the commoner sizes of bolts.

The head of the bolt is chamfered or bevelled at 30° as shown, which puts an arc on the end of each flat when three faces are drawn. In this view, the upper one of Fig. 257(a), a radius equal to the diameter of the bolt D gives an arc sufficiently accurate for the central flat; the smaller radius for the adjacent arcs of the two flats is usually obtained by trial and error, or a radius gauge or template is used. However, the principle behind this is shown in the enlarged view of a nut at (d), the distance f being marked on the upper and lower edges and an arc then drawn through the respective points. The precise 30° chamfer is usually omitted on all but large nuts and bolts since it virtually coincides with the small arc.

Fig 257(b) gives the standard proportions for M10 nuts, which are slightly thicker than the head of the corresponding bolt. Metric nuts are usually chamfered on both faces, but one can be a full bearing face as at (c). On all bearing faces, the nut is slightly countersunk at an included angle of 120° for the depth of the thread as shown, so that the thin end of the thread is not pulled out beyond the bearing surface.

A flat ring *washer* is often inserted beneath a bolt head or nut (see Fig. 258(c)), either to prevent the corners of the head or nut from digging into the surface of the part or to overcome any roughness or unevenness of that surface.

In the United Kingdom, several standard screw threads have been in use for many years and they are generally different in form or profile. The commonest is the British Standard Whitworth thread (BSW), but there are also British Standard Fine (BSF), British Standard Pipe (BSP), British Association (BA) and the Unified thread (UN). BSF is a thread of Whitworth profile but has finer threads than BSW. BSP is, as its name implies, used in pipework. BA thread is used for small screws up to a maximum

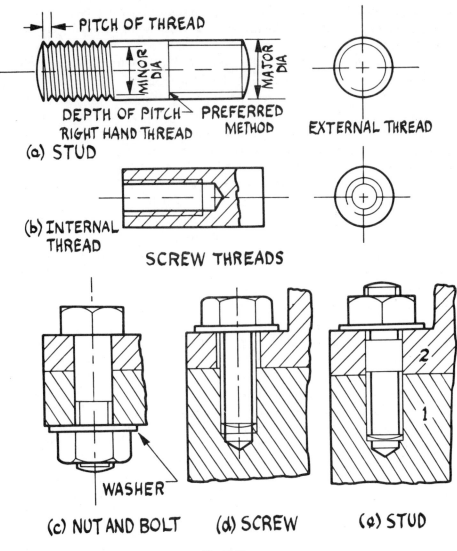

Fig. 258

diameter of about ¼ in. Unified thread is a more modern thread with the same profile as the metric thread but based on inch units. Most of these British threads are due to be supplanted by metric threads but will no doubt remain in use for some years.

Since BSW screw fastenings are so common, the proportions of a standard $\frac{1}{2}$ in. BSW bolt and nut are given in Fig. 257(e). Similar proportions may be used for other BSW bolt and nut sizes for the purposes of engineering drawing. It will be noticed that the thicknesses of bolt heads and nuts are slightly greater for BSW fastenings than for metric ones, and the student is advised to memorize the proportions of both metric and BSW hexagon head bolts, screws and nuts.

A *stud*, Fig. 258(a) and (e), has no head but is simply a straight shank threaded at each end. Normally, one of the parts to be joined has a tapped hole to receive one end of the stud and the other part has a clearance hole to fit over the plain shank of the stud. The stud is screwed into the first part, item (1) in Fig. 258(e), the second part (2) is fitted over the plain shank and a nut is then screwed on the free end of the stud to secure the two parts together, as shown in the figure. Studs are commonly used to secure parts to a main block or casting, e.g. to secure the cylinder head and other parts to the engine of a motor-car.

The main difference between a *screw*, Fig. 258(d), and a bolt is that a screw is usually threaded the full length of the shank, the thread being carried as near to the head as possible. There are many types of screw head, e.g. hexagon, square, countersunk, round, cheesehead, etc.

The *pitch* of a thread, Fig. 258(a), is the distance from the centreline of one thread to the centreline of the next and is thus the distance the nut travels along the bolt during one turn, while the diameter of the screw over the tops of the threads is the *major diameter* and that over the roots is the *minor diameter*.

Normally the thread of a bolt or screw is never drawn in detail on an engineering drawing, but a *conventional representation* of a second line parallel to the shank is used, as shown in Figs. 257 and 258. In such conventional representations, the depth of thread indicated by the inner line need not be precise and a minor diameter of 0·75 to 0·8 of the major diameter is sufficiently accurate. Compare particularly the external and internal threads of Fig. 258(a) and (b) as they should be indicated on the drawing.

There are several other common forms of screw threads, such as square threads, Acme threads, buttress threads, etc., but these are all used for transmitting motion, e.g., moving a table or slide on a machine, and are not used as fastenings since they would not ' hold ' tight.

Screw threads are discussed in further detail in Chapter 3 of Book 3 of this series.

Figure 259(b) shows two *grub screws*. These are headless screws with a shaped end and provided with a slot for fixing. They are threaded throughout their length.

Fig. 259(c) shows two common *keys* used in engineering. They are called keys because they are used to lock pulleys and wheels in position on rounded shafts. A slot is made in both the shaft and pulley or wheel and the key is driven into position when the slots are aligned. The gibhead key is used to that it can be withdrawn more easily.

For heavy and variable loads, and particularly where the wheel may need to be moved along the shaft (axial movement), a *splined shaft* is often used (Fig. 259(d)). The shaft carries a number of splines, as shown, and the hub of the wheel is correspondingly splined to fit them. Splined shafts are used frequently on cars, and on some mechanical toys. (The above items in Fig. 259 are discussed in more detail in Book 3.)

Two *permanent fastenings* are shown at (e) and (f).

Rivet heads can be of several types, but the commonest ones are shown in the sectioned diagram. Large rivets are usually heated until red hot, placed in the common hole and sealed into shape by machine or by hand.

Much riveting is now being superseded by gas and electric *welding*. For interest, Fig. 259(f) shows two simple electrically welded joints. The method is carried out by means of an electrode and a strong electric current, which melts the steel core of the electrode and fuses it into the metal of the parts to be joined.

Once again, these items are discussed at greater length in Book 3.

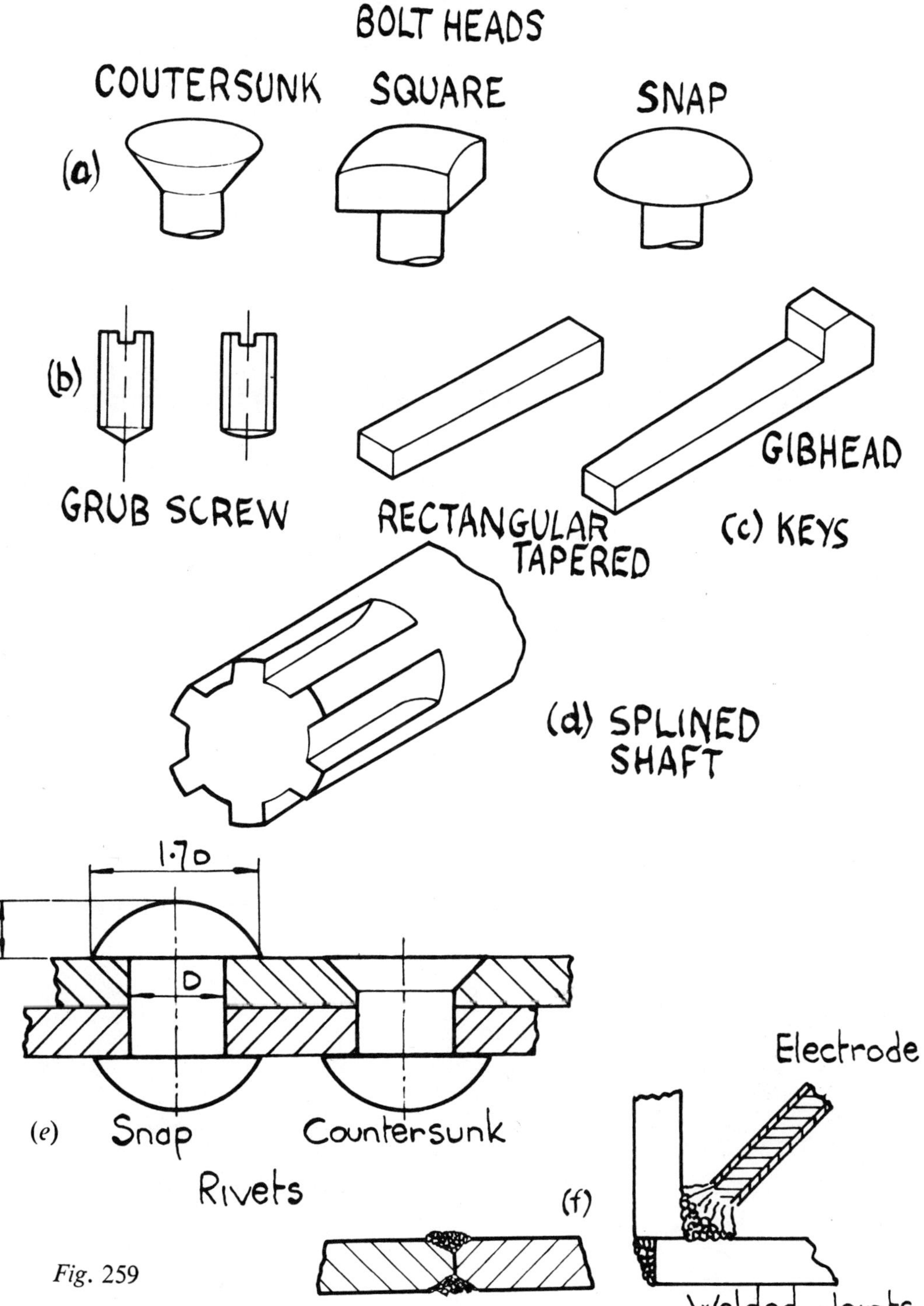
BOLT HEADS
COUTERSUNK
SQUARE
SNAP
(a)
(b)
GRUB SCREW
RECTANGULAR TAPERED
GIBHEAD
(c) KEYS
(d) SPLINED SHAFT
1·7D
·6D
D
(e)
Snap
Countersunk
Rivets
(f)
Electrode
Welded Joints

Fig. 259

CHAPTER TWO

Freehand Sketching

The ability to make a clear sketch of an object is of great importance to a draughtsman, for many working drawings made by him begin as freehand sketches. Also a freehand drawing, made only with a pencil, forms a compulsory part of many examinations.

Freehand sketches can be drawn orthographically of isometrically. The method used sometimes depends on the shape of the object. It would be easier, for instance, to make an orthographic sketch of a spanner, rather than isometric one.

In any case, orthographic views should be drawn in the early stages. If the student begins with simple objects and practises regularly, his efforts will soon become quite presentable.

Freehand sketching need not be drawn to scale, but it must be drawn in good proportion. It is good practice to begin with a 25 mm square, to measure it afterwards and note if it is fairly accurate. If there is a considerable difference, the exercise should be tried again.

When this is satisfactory, the three views of a rectangular block could be drawn, say 50 mm long, and of width and height equal to one half and one third respectively of the length. Again the final result should be checked by measurement.

From this stage, familiar objects can be sketched, choosing those of increasing difficulty.

Some useful points to bear in mind are:

1. Hold the pencil (the H one should always be used) lightly, so that freedom of movement is obtained.
2. When copying an object, study the original carefully before deciding on the position in which to show it.
3. Start with the view giving the clearest idea of its shape, usually the front elevation.
4. Begin with centrelines whenever possible, and outline the object evenly about them.

5.   Sketch lightly until the views are finished.   Then line in.
6.   Add all necessary dimensions and notes.

It is good practice to study an object for several minutes and then sketch it as accurately as possible from memory.

Fig. 260, a mallet-head, shows how to carry out the above rules in two stages. Begin with the centrelines and build up a faint outline around them, as in the left-hand drawing. When this is a good likeness, clean up the sketch, line it in and dimension it, as shown on the right-hand side.

Here are a few sketches to attempt, graded in difficulty:

A wooden wedge	A stool
A housed joint	A small table
A bridle joint	An M20 bolt and nut
A wooden ink-pot stand	A kettle
A simple book-end	A pair of pincers
	A small vice

An advantage of isometric sketching, which can be developed in a similar way, is that only one view is necessary to give a good idea of the object.

Begin by sketching a 40 mm isometric square, then check the 30° axis for accuracy. If necessary, redraw the square until it is reasonably accurate. Next an isometric cube can be attempted, then the rectangular block already drawn orthographically.

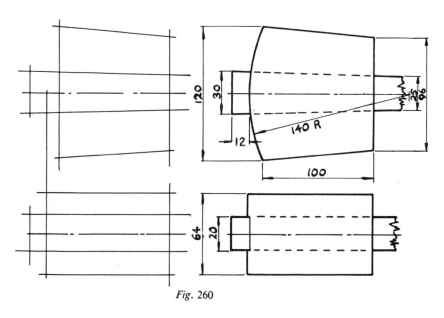

Fig. 260

When this has been done satisfactorily, shaped objects such as the bracket in Fig. 261 can be attempted. Note that the isometric crate is first drawn as described in Part II.

Avoid, if possible, isometric sketches of objects with curved surfaces, but if these are required the principle of the American method (Fig. 262) will give a good freehand reproduction.

Sketch the isometric square and mark on it the mid-points A, B, C, D. Draw the diagonal EF and the lines GB and GA. Mark points J and K approximately equal to HB. When the paper is turned so that EF is horizontal, it should be possible to draw a fair isometric circle through the points marked.

Fig. 263 is a freehand application of the method of Fig. 126, the length of the cylinder being marked along the projectors and the resulting points joined with a parallel curve. Book 3 contains many further freehand exercises.

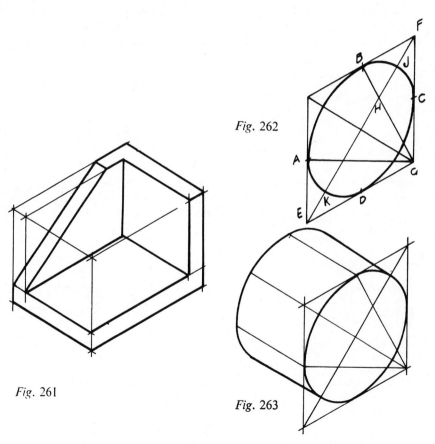

Fig. 262

Fig. 261

Fig. 263

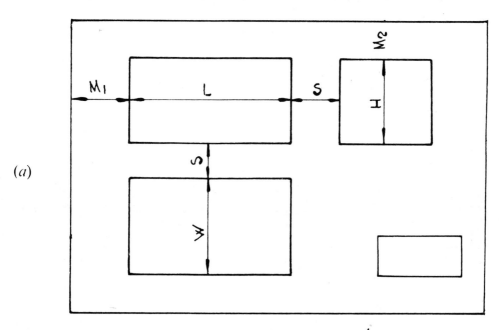

$$M_1 = \frac{\text{Length of Paper} - (L+S+W)}{2}$$

Fig. 264

$$M_2 = \frac{\text{Width of Paper} - (H+S+W)}{2}$$

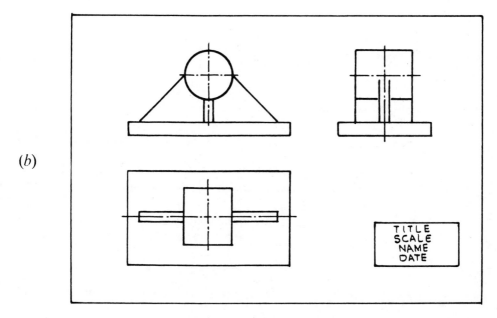

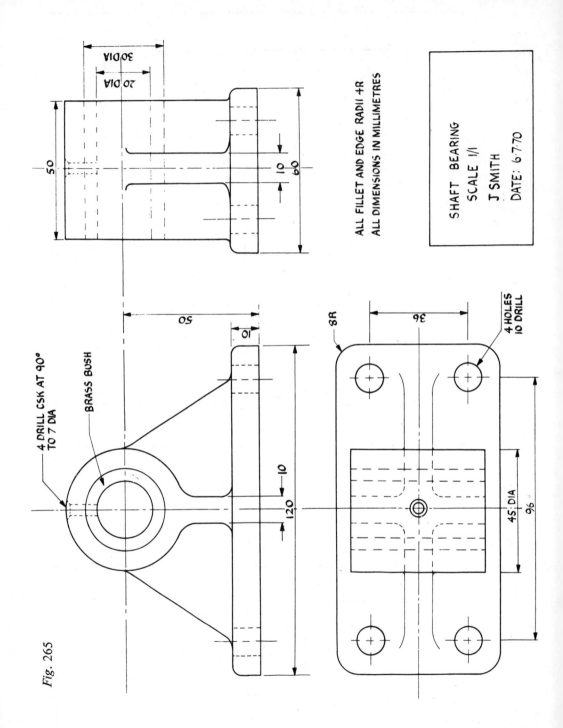

Fig. 265

CHAPTER THREE

Making the Drawing

A. THE LAYOUT

Before attempting an engineering drawing it is most important to study the original drawing set as an exercise, so that the work involved is fully realized. This will be discussed later.

The next point is to set out the drawing correctly on the paper. This can be done by first reading the overall dimensions of the object; that is, the full length, width and height.

Fig. 264(*a*) shows the rest of the calculation, and you are advised to work out the position of your views in this way before you begin. Valuable time and marks are often lost by students who do not do this, for after proceeding for some while, they find that they have not left sufficient space to keep the views in projection.

Fig. 264(*b*) shows the layout of the drawing, following the procedure of Fig. 264(*a*). Notice how balanced the drawing appears on the paper, and therefore more attractive. Dimension S will vary according to the size of the drawing, but it should not be less than 15 mm, and 25 mm for most objects will be a reasonable allowance. Also, observe the use for the small rectangle. It is called the label, and should carry the information indicated, including the title, no matter how small or large the drawing.

Fig. 265 is a full-size drawing of the object outlined in the previous figure. Except for the distances between the views, which have necessarily been limited by the size of the page, it is set as an example for study.

Note the following points:
1. The position of the three views. They must be in this order, always, in first angle projection. (For exceptions see page 115.)
2. The views are in projection, drawn around two centre-lines at right angles to each other.
3. Projection lines and a ground line are not shown on this finished drawing, which makes the object clearer.

4. Hidden edges are shown by means of broken lines.
5. Dimensions:
A complete arrow or *dimension line* (not broken) is used, and the dimension is written above horizontal arrows and to the left of vertical ones. Study the other points below carefully.
   (a) All dimensions are written at right angles to the dimension lines.
   (b) All arrow heads are small, neat and blocked in.
   (c) There is a small gap between the object and the *witness* or *extension lines*.
   (d) Dimensions are normally taken well clear of the object. Where this is not possible, they are written in clear spaces on the object.
   (e) Shorter dimensions are put nearer the object, longer dimensions outside them.
   (f) Centrelines are not used as dimension lines, though they may be used as extension lines.
   (g) The *radii* of arcs are given, the arrow head at the centre being omitted, and the abbreviation RAD or R is used, but the *diameters* of circles are given (two arrow heads being used), and the abbreviation DIA. or ϕ. is added.
   (h) All printing and figures are easily readable.

All the above points must be borne in mind in engineering drawing, and each student should take pride in his draughtsmanship so that his work is distinctive.

(Note: The *webs* shown in Fig. 265 are metal ribs or partitions which act as a connection or support. They occur frequently in the examples which follow, and also in examinations.)

B. READING THE DRAWING

As mentioned at the beginning of this chapter, when given a drawing as an exercise you must study the views until a clear picture of the object is formed in your mind. This may be a little difficult at first, but will soon come with practice. Remember the relationship of the views, as explained in Chapter 2, and train the eye to move horizontally and vertically from one view to another.

If there is any part in one view which is difficult to visualize, such as that shown by hidden lines, relate that part to its position in the other views, using a straight edge if necessary, providing the drawing is in projection.

By studying its shape in the other views, you will soon obtain in your mind the picture you require.

For instance, suppose you are confused by the broken lines shown in the plan of Fig. 265. By referring to the elevation the outer lines will quickly be understood as the sides of the brass bush—a metal lining which can be changed when worn—seen from above. The other broken lines will then be seen as continuations of the webs.

The meaning of the broken lines in the end elevation can be found in a similar way, and a complete picture of the bearing can then be formed. When this is achieved, you can say that you are able to ' read the drawing '.

Where a curved face merges into a straight face or another curved face, no line is shown on the face view, e.g. where the lower outline of the boss meets the web in the end elevation of Fig. 265. The junction is usually outlined in a related view—in this case, in the elevation of Fig. 265, where the web is shown to run into the boss by a *fillet* or curve of 5 mm radius.

The outlines of the four webs in the plan are, of course, their vertical edges as seen from above. The only indication in the plan that the webs parallel to the V.P. are rounded at their base, is given by the arcs at each end. But by referring to the end elevation this fact is made quite clear, and it can also be seen that the web runs into the boss by means of a similar curve.

In the same way, by referring to the elevation, it can be seen that the webs at right angles to the V.P. are likewise rounded to the base and the boss.

Such a careful reference to all views is an essential part of the ability to read a drawing correctly.

C. THE METHOD

The method of building up the drawing is shown as simply as possible in Figs. 266(*a*), (*b*) and (*c*).

Until the final stage it is good practice to draw all lines lightly, using only slightly more than the weight of the pencil. When mistakes are made they can then be easily erased without spoiling the final appearance.

(*a*)   The layout is decided upon and the centrelines are drawn to agree with this arrangement. The three views should be built upon these centrelines.

(*b*)   The chief parts of the object are located. In this case, the centre of the boss (the cylindrical part) is the key position. The outline of the boss is drawn and projected to the other views. The positions of the webs can now be located, and the base plate drawn in all views.

(c) Further details are now added. The elevation of the bush is included and its hidden edges are drawn in the outer views. The hidden edges, also, of the webs are marked. The positions of the four holes are found, and the centres of the arcs for all corners.

After the drawing has been carefully checked, it should be lined in. All the unnecessary construction lines are erased, corners cleaned off and finger and other marks removed. When lining in, draw all circles and arcs first, for it is easier to draw the straight lines to meet them than the reverse.

Finally, the drawing will be correctly dimensioned, a label added, and the usual information filled in with neat printing. A frame can then be added round the drawing, if desired, to give the work a ' professional ' appearance.

Following these instructions carefully, the student should be able to make a reasonable copy of Fig. 265.

Further exercises, graded in difficulty, are given below.

EXERCISE 1

(Note: Include an appropriate title and the scale on all drawings attempted.)

1. Fig. 267(a) shows an isometric drawing of an *end bracket*. Draw full size a front view as seen in the direction of arrow 'C', and an end view and plan in projection with this. Insert two horizontal and two vertical dimensions on the drawing.

2. An *angle bracket* is shown in Fig. 267(b). Draw an elevation looking in the direction of arrow 'D', an end elevation and plan in first angle projection, showing hidden lines.
   Scale: full size. Insert four dimensions.

3. Fig. 268 gives details of a *slide block*. Draw an elevation in direction of arrow 'E', and end elevation and plan of the block, full size and showing hidden detail. Include four dimensions.

4. Details of an *open bearing* are given in Fig 269. Draw full size an elevation as seen in direction of arrow 'F', and an elevation and plan in projection. Show all hidden detail and include six important dimensions, three horizontal and three vertical.

5. Two elevations of a metal slide block are shown in Fig. 270. Draw full size the following views: (a) The given elevation ' E '; (b) An elevation in the direction of arrow F; (c) A plan in the direction of arrow P. Fully dimension elevation (b).

(a)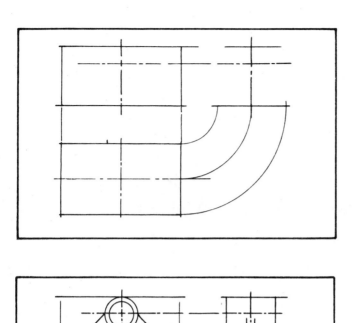

Layout decided upon;
Centre lines drawn

Chief parts located;
Webs and base plate drawn

(b)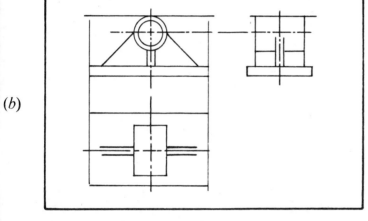

Further details added and outline completed

(c)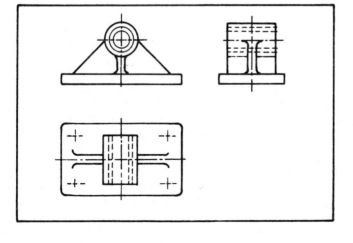

Drawing to be cleaned up, lined in and dimensioned

Fig. 266

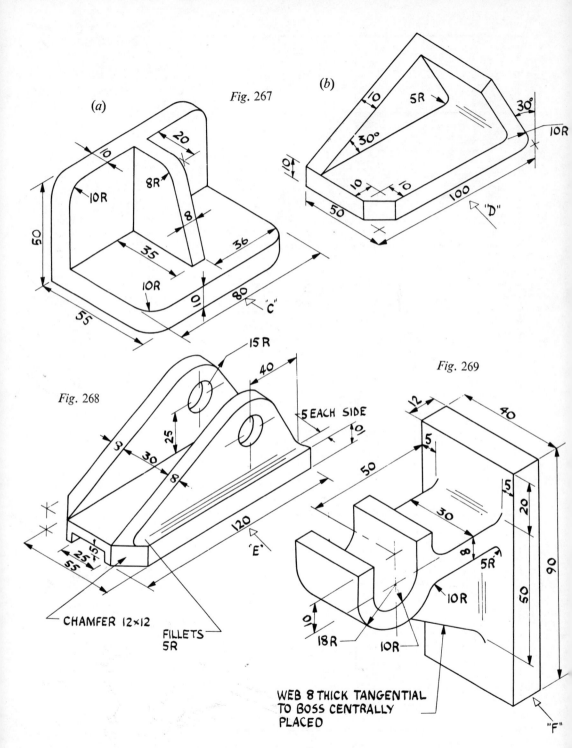

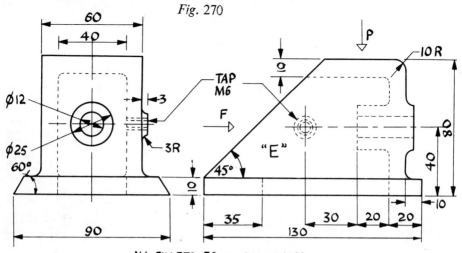

Fig. 270

ALL FILLETS 5R UNLESS STATED

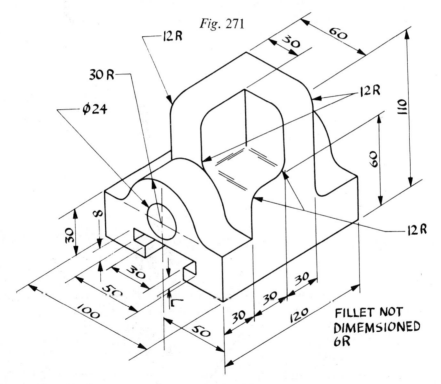

Fig. 271

6. An isometric sketch is given of a jaw support (Fig. 271). Draw full size:
   (a) Front elevation.
   (b) Side elevation.
   (c) Plan.

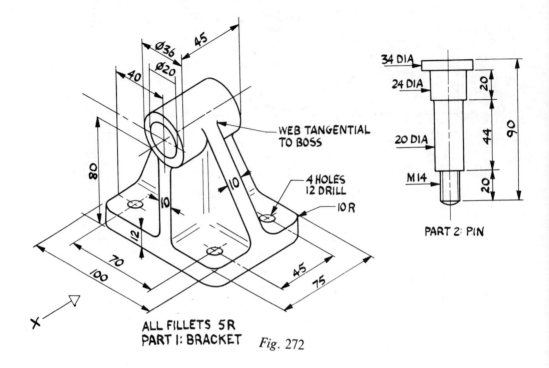

Fig. 272

7. A cast iron bracket and a pin are shown in Fig. 272.. The pin is to be fitted into the boss and secured with a washer and nut.
   Draw full size the following views of an assembled bracket and pin:
   (a) An elevation in the direction of arrow 'X'.
   (b) An end elevation.
   (c) A plan projected from drawing (a).
8. Simplified details of the cover arrangement for a small valve are shown in Fig. 273. The bridge is attached to the cover by two M8 bolts and the spindle is screwed through the cover into the central hole in the bridge.
   Draw full size the following views of the assembled cover arrangement:
   (a) An elevation.
   (b) A plan in projection with your elevation.
   (c) An end view in projection with the elevation (a).
   (Note: Assembly drawings are discussed in detail in Book 3, where further examples are given. In Fig. 273 observe the symbol ϕ. This is a modern method of indicating diameter and often appears in examinations. It always *precedes* the dimension, e.g., ϕ 24, not 24 ϕ.)

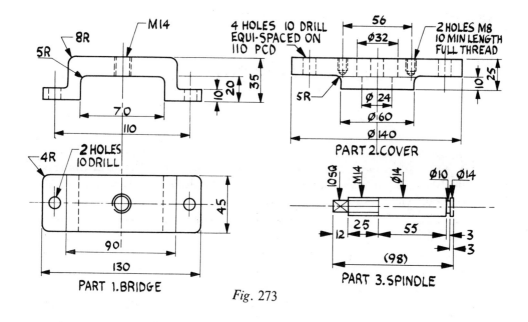

Fig. 273

CHAPTER FOUR

Sectional Elevations and Plans

Section planes and sectional views were discussed in detail in Chapter 5 of Part II, so in this chapter all that need be dealt with is how to present sectional views of engineering parts.

Study Fig. 274. This shows an elevation and end elevation of a pedestal bearing bracket. The elevation is really two views, for the half to the left of the centreline is an outside view, and the part to the right of the centreline, an inside view; that is, a sectional elevation through B–B in the end elevation, looking in the direction of the arrow.

The end elevation has also been sectioned. It is really a view looking through the centreline A–A of the elevation, in the direction of the arrows, and is thus on the *left* of the elevation.

It must be remembered that whenever an end view is required as seen from the *right* in first angle projection, it must be drawn on the *left* of the elevation.

There are other interesting points to be observed. The bush is of different metal from the rest of the bracket. This is shown in section by drawing hatched lines in opposite directions to those of the bracket. Notice, too, that the lines are drawn slightly closer than the other hatched lines to emphasize the difference. Furthermore, the lines in the bush itself run in opposite directions, and when the half-sectional elevation is studied more closely, the reason for this will be seen. The bush is in two parts, and this is the method of showing it clearly.

It will now be realized that the bracket itself is in two parts, for it is likewise hatched in opposite directions. Note, however, that the hatching does not cover the vertical hole.

Hatched lines must never cross full lines, for if the component is really in two parts, the parts must be hatched in different directions as shown. One exception to this is a section showing an internal screw thread, where the hatching crosses the outer line to the inner line of the thread (see Fig. 258 (b).

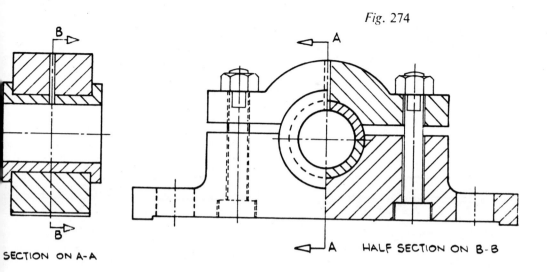

Fig. 274

SECTION ON A-A   HALF SECTION ON B-B

Plain cross-hatching is now used for most materials, although on some old drawings different materials are marked in different ways. Nowadays, such information is given in notes on the drawing to simplify the work involved, or different colours are used for identification.

What of the bolt in the half-sectional elevation? This is shown in outline because no useful purpose is served by showing it otherwise. Several similar parts are not shown hatched when a section plane runs through them parallel to the main shaft.

They include: shafts—such as that in Fig. 275(c)—screws, studs, keys and webs. These must, therefore, always be left in outline when a cutting plane is presumed to 'penetrate' them.

However, in the case of shafts and webs there are exceptions. Figs. 275(a) and (b) show two views of a simple bearing bracket and shaft supported by four webs on a base plate. If a section is taken through B–B, the sectional elevation will be as at (c). The webs parallel to the section plane are not shown in section, neither is the shaft, except at its broken ends, which are drawn in this way to indicate that the shaft has been shortened.

The vertical portion shown hatched at (c) is not a section of the web at right angles to the cutting plane B–B, though it may seem to be at first. It is a section of the body of the bearing, joining the boss to the base. A study of (e), which is a horizontal section on C–C, may make this clearer. Here, the imaginary inner limits of the webs have been drawn in, and the resulting square in the centre, shown more finely shaded, is a part of the main object.

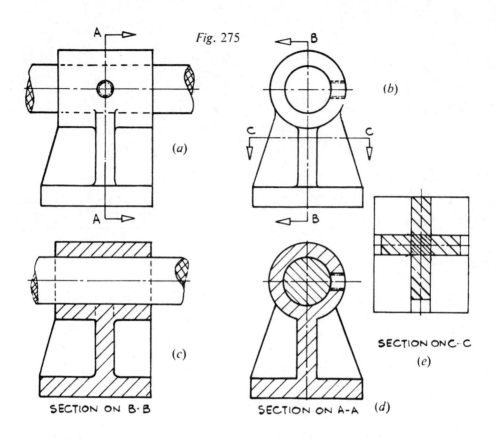

Fig. 275

SECTION ON B-B (c)

SECTION ON A-A (d)

SECTION ON C-C (e)

(d) shows a sectional end elevation on A–A. Notice that the other two webs are now parallel to the section plane, and are therefore shown in outline, while the vertical portion of the bearing remains hatched as in (c). The shaft is cut across at right angles to A–A and must be shown hatched in this view.

A rule for guidance is:

When a web or shaft is parallel to the section plane, it is not *shown in section. When it is at right angles to the section plane (that is, cut across by the plane), it is shown in section and must be hatched.*

Two other useful points can be learnt from (d).

The small hole shown in the elevation is a tapped hole. When such a threaded hole is penetrated by a section plane, it is shown by a double line as at (d), and the hatching lines cross to the inner line, as mentioned above.

When hatching with the set-square across an irregular section, as in (c) and (d), look for all portions to be hatched and take care that corresponding hatch lines across the different portions are done at the same time, and thus kept in line. The hatch lines will then coincide when they meet on a common portion. Thus, in (c), the hatching starts with the boss and the set-square soon begins to cross the base; corresponding hatch lines are drawn on boss and base and coincide when the central web is reached. If this procedure is not followed, the hatch lines of boss, web and base might not meet one another, nor preserve a regular and attractive pattern.

Finally, sectional views should be projected in the direction shown by the sectional arrowheads referring to them (see Fig. 274).

In Question 1 below, the required end elevation should be drawn to the left of the front view. Furthermore, the required plan should be drawn *above* the elevation in this instance with the section of the projecting boss resting on the section of the base.

These particular cases are the only exceptions to Rule 3 on page 8, but they must be remembered if first angle projection is to be consistent on the drawing.

From all this, it will be appreciated that it is difficult to draw a sectional view unless every aspect of the drawing under consideration is thoroughly understood.

Make a copy of the four views of Fig. 275.

When this is completed, make a copy, twice the size if possible, of the two views of Fig. 274.

EXERCISE 2

1. Fig. 276 shows two views of a pipe guide. Draw these views full size, and project a sectional plan on A-A and a sectional end elevation on B-B.

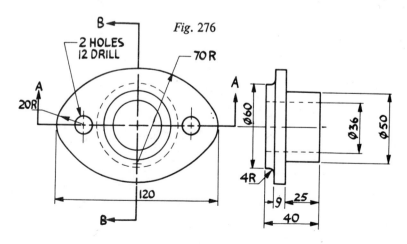

2. Two views of a bracket are shown in Fig. 277. In the place of the given views draw a sectional elevation on B-B and a sectional end elevation on AA, both full-size.
3. Fig. 278 gives two views of a mild steel boss fixed to a rectangular base plate. In the place of the elevation given, draw a sectional elevation on B-B, then project a sectional plan on A-A, and a sectional end view on C-C.
4. Two views of a pump casting are given in Fig. 279. Draw, full size, the following views:
   (a) A sectional elevation on A-A.
   (b) To the left of the above view, a sectional elevation on B-B.
   (c) In projection with view (a), a plan showing all hidden detail.
   Small radii may be drawn freehand. Dimension the plan only. Add the title PUMP CASTING and also the scale.
5. Fig. 280 shows two views of a machine casting. Draw the following views full size:
   (a) The elevation shown.
   (b) A sectional end elevation on A-A.
   (c) A sectional plan on B-B.
   Dimension drawing (c). Add the title MACHINE CASTING and also the scale.

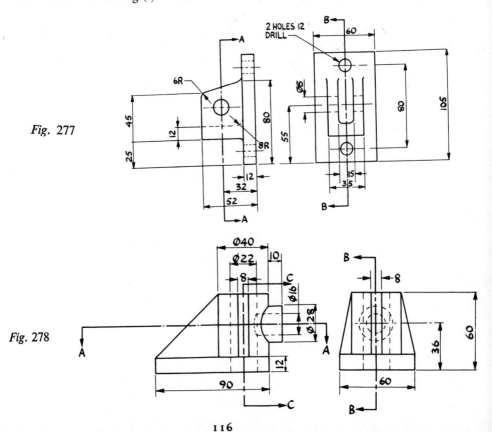

Fig. 277

Fig. 278

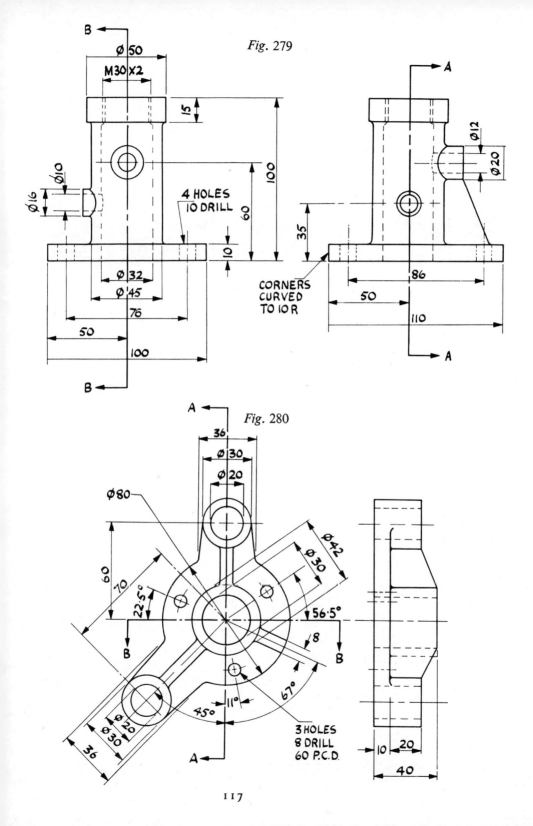

Fig. 281

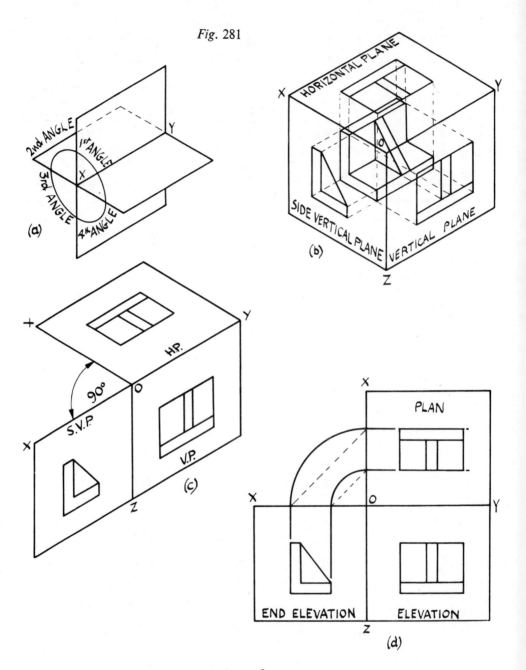

CHAPTER FIVE

Third Angle Projection

In Orthographic Projection space is imagined to be divided by two principal planes (this does not include the side vertical one) intersecting at right angles as shown in Fig. 281(a). The four angles produced by this division are called quadrants or, more technically, *dihedral angles*. The student, already familiar with the first angle, should now remember the position of the other three angles, particularly noting that they are numbered in an anti-clockwise direction.

The second and fourth angles of projection are rarely, if ever, used in engineering because to do so might lead to confusion, but third angle projection is becoming increasingly popular and therefore merits consideration.

Study Fig. 281(b) for some moments. At first it may appear to be a shapeless pattern of intersecting lines, but if the centre is concentrated upon, an isometric drawing of a square bracket stiffened with a central triangular web may gradually emerge.

This bracket is suspended in space and around it are arranged the reference planes of the third dihedral angle. Note that the planes come between the observer and the object, for this is the essential difference with this method of projection. While in the first angle the object is between the planes and the observer (see Fig. 102), in the third angle it is the *planes* which are nearer the observer and between him and the object.

To overcome the difficulty which this arrangement presents, the planes are assumed to be transparent, as are any auxiliary planes which might be used in addition, and the views are merely those seen through the planes but, as it were, reflected back on them. The broken lines, some of which have been omitted for clarity, indicate this 'reflection'.

With each view 'impressed' on its respective plane rebatement is now in theory carried out. The side vertical plane (S.V.P.) with its impression is rotated through 90° as shown in Fig. 281(c) so that it is common with the vertical plane (V.P.). These two planes are then rotated up through 90° to meet the horizontal plane (H.P.) so that all three views are now in the one plane.

This is the principle of projection in the third angle and it will be seen that the three views of Fig. 281(d) are in exactly the opposite relationship as they would appear in the first angle, for the plan is now above the elevation and the end elevation to the left of the front view.

The main advantage of this arrangement is that each view shown represents the nearest face in the adjacent view. For instance, the elevation is really the view represented by the lower horizontal line of the plan, and the end view is that represented by the extreme left-hand vertical line in the elevation. This should be read through carefully again with reference to Fig. 281 until it is fully understood.

Apart from this fact, why third angle projection?

It is used almost exclusively in the U.S.A. (it is often called 'American Projection') and Canada, and has been adopted by the Armed Forces of Great Britain and, for these reasons, by most leading engineering firms. First angle projection (otherwise 'English Projection') still prevails in some firms, especially those exporting to certain countries in Europe, but the time may come when American Projection is used exclusively here also.

The student should not worry if the method at first seems difficult to put into practice, because after a few exercises it will soon become familiar. Furthermore, first angle projection is still accepted in all examinations unless specifically stated to the contrary, mostly because it follows in the earlier traditions of formal geometry.

The following points should be borne in mind:
1. In problems on solid geometry adhere whenever possible to first angle projection.
2. In technical drawing exercises draw the views projected in the angle asked for—use the first angle if none is mentioned.
3. Indicate which angle has been adopted, particularly if the third angle is used.
4. Keep to one method on each drawing. In other words, never use first angle and third angle together in the same question.

Fig. 282(a) shows a simpler method of setting out views in third angle by dispensing with an XY line. Begin with the front elevation, building up the end elevation during the construction. The horizontal phantom line joining the top of these views becomes a reference line for locating the plan, all points on the end elevation being projected vertically to this line before being taken across at 45° as shown. Alternatively using point ' O ', the distances can be radiated with the compasses.

Fig. 282(a)

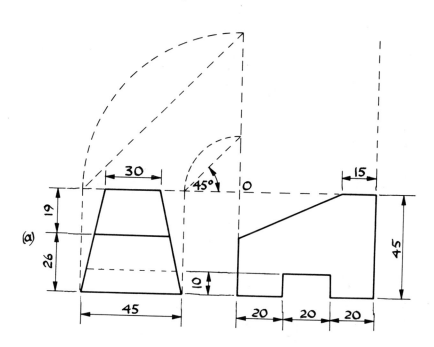

All the exercises which follow are required to be drawn in third angle projection.

1. Copy the two views of Fig. 282(a) full size. Then project a plan of the block by one of the methods shown.

2. A pictorial view of a tube hanger is shown in Fig. 282(b). Draw an elevation as seen in the direction of arrow 'A', and a plan and end view in projection with this.
3. Fig. 282(c) is an isometric drawing of a slide valve. Draw a front elevation as seen in the direction of arrow 'B', then a plan and elevation projected from this.
4. Fig. 283 gives the elevation of a crank lever with two auxiliary sectional views. Draw three views of the lever in American projection.

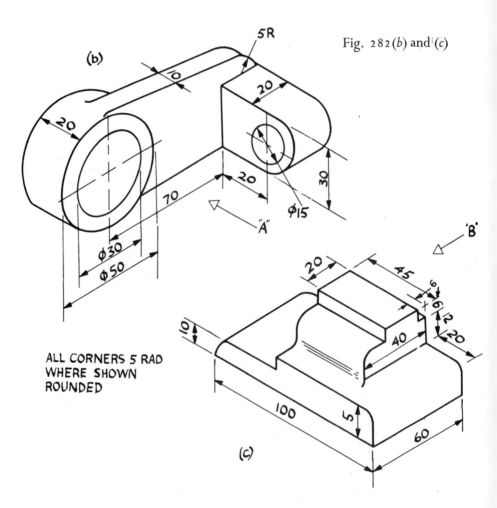

Fig. 282(b) and (c)

ALL CORNERS 5 RAD WHERE SHOWN ROUNDED

5. Two views of a bearing bracket are shown in first angle projection in Fig. 284. Draw full size three views of the bearing in third angle projection. Insert six dimensions, three horizontal and three vertical ones, and in a label include an appropriate title, the scale of the drawing and your name.
(Note: In exercises of this kind, decide on the front elevation in the given figure and copy this first, since it will also be the front elevation in third angle.)

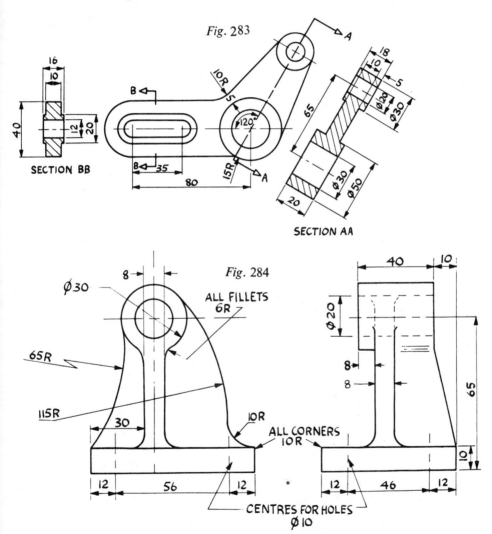

Fig. 283

Fig. 284

FIRST ANGLE PROJECTION

Test Papers

These tests are taken from G.C.E. examination papers set by the Examining Boards shown. About an hour and a half is a reasonable time allowance.

TEST 1

Fig. 285 shows details of a footstep bearing for a vertical shaft of a machine. The bearing is attached to the underside of the machine by the four tap bolts.

With the parts correctly assembled draw, *full-size*, the following views:

(a) a sectional front elevation taken on the centre-line and looking in the direction shown by the arrows X–X,

(b) an outside end elevation looking in the direction shown by the arrow ' Y ',

(c) an outside plan looking in the direction shown by the arrow ' Z '.

Hidden detail to be shown in view (b) *only*.

Only ONE tap-bolt is to be shown, that for hole H marked in the plan of the body. The bolt, including all its hidden details, must be shown in all three views.

Insert the following dimensions:

(i) The distance between the centres of the holes for the tap-bolts in both directions.

(ii) the internal diameter of the bush.

(iii) the overall length of the bush.

(iv) the vertical distance from the top of the spherical seating to the top face of the bolt flanges.

In a rectangle 150 mm long and 75 mm high in one corner of the drawing insert, in letters 10 mm high the title FOOTSTEP BEARING and, in letters 5 mm high, the scale and system of projection used.

First-angle or third-angle projection may be used but the three views must be in a consistent system of projection. (A.E.B.)

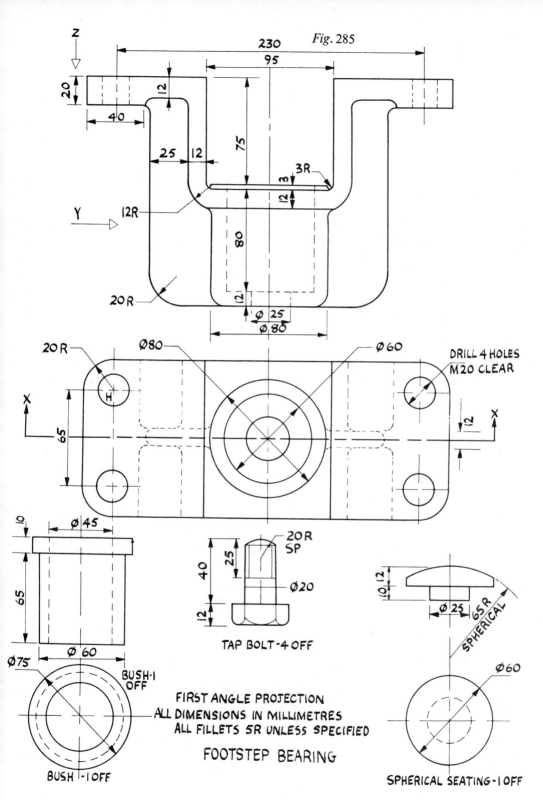

TEST 2

Fig. 286 shows details in first-angle projection of a cylinder cover and stuffing box for a small air pump.

With the parts correctly assembled and the gland inserted in the cover of 5 mm, draw, *full size*, the following views:

(a) a front elevation looking in the direction of arrow ' A ';
(b) a sectional end elevation, the section being taken through the centre of the cover and looking in the direction of arrows B–B;
(c) a plan looking in the direction of arrow ' C '.

In the sectional view (b) show a length of approximately 150 mm of the pump rod, 25 mm diameter, passing through the stuffing box. Hidden details are not to be shown.

Insert the following dimensions:
  (i) the diameter of the pump rod,
  (ii) the diameter of the spigot on the cover,
  (iii) the pitch circle diameter of the holes for the studs attaching the cover to the cylinder,
  (iv) the diameter of the studs for the gland,
  (v) the centres of the studs for the gland.

In a rectangle 150 mm long and 75 mm high in a corner of the drawing insert, in letters 10 mm high, the title STUFFING BOX and, in letters 5 mm high, the scale and system of projection used.

First-angle or third-angle projection may be used but the three views must be in a consistent system of projection. (A.E.B.)

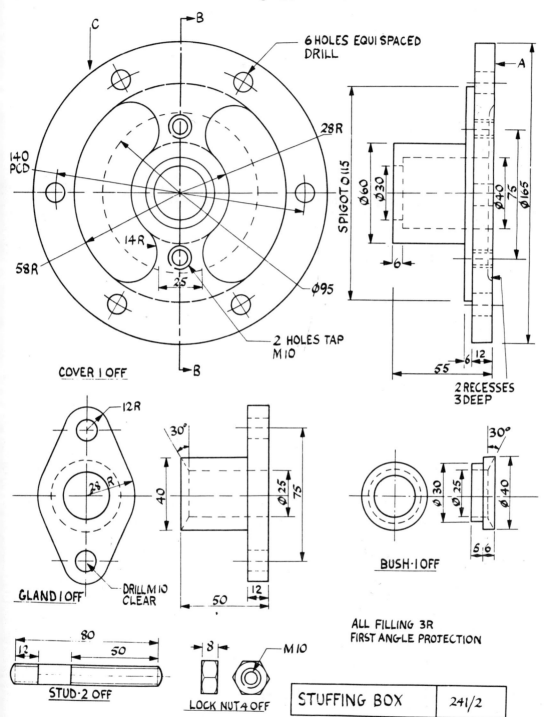

Fig. 286

STUFFING BOX 241/2

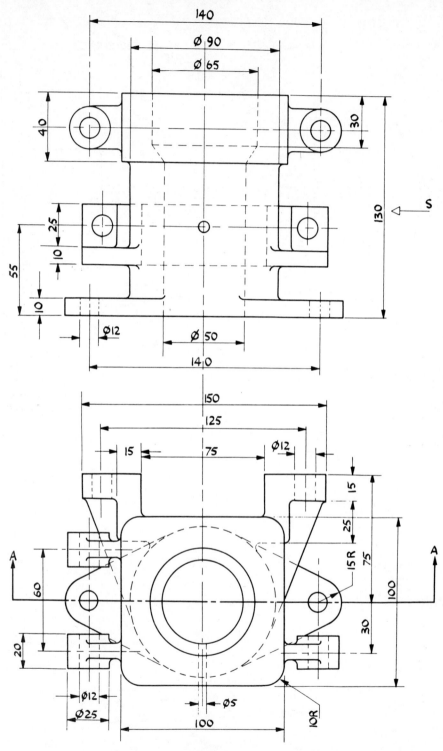

Fig. 287

TEST 3

Fig. 287 shows, in first-angle projection, two views of a cylinder cap.

Draw, full size and in correct orthographic projection, the following views:

(a) a sectional elevation on A–A looking in the direction of the arrows;
(b) a plan;
(c) an elevation looking in the direction of the arrow 'S'.

Hidden edges are not to be shown on any of the views and dimensions are not required.

Either first-angle or third-angle methods of projection may be used; the method chosen must be stated on your drawings.

In the bottom right-hand corner of your paper draw a title block 100 mm by 60 mm and in it print neatly the drawing title, the scale and your name.

L. U.

Some Commoner Abbreviations Used on Drawings

British Association	B.A.	Not to scale	N.T.S.
British Standard	B.S.	Outside diameter	O/D
British Standard Whitworth	B.S.W.	Pitch circle diameter	P.C.D.
		Radius	RAD or R
Centreline	CL.	Right hand	RH.
Centres	CRS.	Round head	RD. HD.
Counterbore	C'BORE	Spotface	S'FACE
Countersunk	CSK.	Standard	STD.
Diameter	DIA or ϕ	Undercut	U'CUT
Internal diameter	I/D	Système International	SI
Left hand	L.H.	Threads per inch	
Machined	M/CD	(B.A., B.S., B.S.W.)	T.P.I.
Metric	M		

Materials

Cast iron	C.I.	Aluminium	Al.
Wrought iron	W.I.	Brass	Br.
Mild steel	M.S.	Copper	Cpr.